Animal Society

Animal Behaviour

Animal Society

Federica Colombo
adapted by Paul-Henri Plantain
translated by Dr. R.D. Martin, F.I. Biol, and A.-E. Martin

Burke Books ▶️ LONDON * TORONTO * NEW YORK

First published in the English language 1981
© Burke Publishing Company Limited 1981
First published in Canada 1981
First published in the United States of America 1981
Translated from *Les animaux en société*
© Vallardi Industrie Grafiche S.p.a. 1979

Library of Congress Catalog Card Number

CIP data
Colombo, Federica
  Animal society. — (Animal behaviour; 4)
  1. Social behavior in animals
  I. Title   II. Les animaux en société, *English*
  III. Series
  591.51   QL775

ISBN 0 222 00823 7

Burke Publishing Company Limited
Pegasus House, 116-120 Golden Lane, London EC1Y OTL, England.
Burke Publishing (Canada) Limited
91 Station Street, Ajax, Ontario, L1S 3H2, Canada.
Burke Publishing Company Inc.
27 Harrison Street, Bridgeport, Connecticut 06604, U.S.A.
Printed by Vallardi Industrie Grafiche S.p.a.. Milan

Acknowledgements

The publishers are grateful to the following copyright-owners for permission to reproduce copyright material:

William Collins & Co. Ltd.
Macdonald & Jane's.
Pan Books Ltd.

The photographs in this book are reproduced by permission of:

Capelli; Bruce Coleman; Marka; Pagani.

The drawings are by Gabriele Pozzi.

Contents

The Behaviour of Animals

The study of animal behaviour, or ethology, is a new science. In fact, it is not as easy as it may seem to define what is meant by "behaviour". The word covers all the activities that an animal performs, but it also includes all the various interactions involved, indeed all the internal and external factors that govern these activities. For this reason, the study of animal behaviour can be approached from two different directions: one physiological and the other psychological.

The physiological approach is concerned with the analysis of the mechanisms involved in responses to sensory stimulation. Physiologists study the way that the nervous system operates during various activities performed by the animal. They examine, so to speak, an animal's behaviour "from the inside" in an attempt to understand why particular activities occur. The psychologist, on the other hand, studies an animal's behaviour "from the outside" by examining the many environmental factors which may evoke particular responses.

Our gradual progress towards a real understanding of patterns of animal behaviour can be compared to a voyage of exploration and discovery. Animal behaviour exhibits an extraordinary degree of diversity, as great as the diversity in shape, size and colour. No two species behave in exactly the same fashion and each animal species can exhibit a very large range of behaviour patterns. For example, in studying the behaviour of seagulls simply with respect to feeding, great variation can be seen. Seagulls may feed on fish caught by diving into the water, on ailing birds which are killed with a stab of the beak, on earthworms which must be dug up from the ground, or on insects caught on the wing. Nevertheless, the range of behaviour exhibited by a particular species has well-defined limits. For example, a gull cannot grasp its victims as a bird of prey does. As yet, the immense diversity of animal behaviour has been described only rather superficially and there are many aspects which are still to be properly studied.

The complexity of the research involved and the enormous variety of the phenomena which can be observed oblige scientists to become very specialized; they can only deal with specific aspects at any one time. Some scientists study the functioning of particular components of their animals, such as the sensory organs, or perhaps restrict their attention to only one sense organ. Others examine the behavioural modifications which have taken place in the course of the evolution of a particular species. There are those who prefer to work in the laboratory using an experimental approach, or to observe their animal subjects directly under natural conditions or in zoological parks. Some scientists specialize on a particular species, some study a group of species living in contact with one another; yet others carry out comparisons involving a large number of species. However, this fragmentation in the work of the scientists

Thanks to their powers of flight, birds have been able to spread to virtually all regions of the world and in the process have become adapted to a great variety of habitats. Such adaptation to the environment has led to the emergence of a wealth of special behaviour patterns which have attracted the attention of scientists. The emperor penguin (left) is pushing its single egg beneath a fold of its belly, which fulfils the function of a nest.

Because of their predominantly nocturnal habits, mammals, such as the red fox vixen (below) in her den with her cubs, are difficult subjects for the study of animal behaviour.

involved is only apparent. Zoologists, physiologists and psychologists are, in fact, all engaged in a common task and they collaborate with one another continually in order to integrate the results of their studies. But, in spite of this extensive collaboration, we are still far from achieving a complete understanding of animal behaviour.

Sensory stimulation is often the starting-point for the performance of a particular behaviour pattern. A dog, seeing its master return after a long absence, leaps in the air and runs to meet him. If the master reaches for the lead, the dog will bark in response to this signal indicating that it will be taken for a walk. On seeing a bone, the dog will wag its tail and sometimes salivate, while the smell of potential prey will cause it to sprint off in the appropriate direction. All these manifestations represent responses made by the dog to aspects of its environment. Ignorance of these basic facts can lead to serious errors in a dog's training. Worse than that, however, is anthropomorphism — that is, seeing the dog's behaviour in human terms. A human infant can, of course, jump for joy on seeing an adult put on a coat prior to going out for a

walk. But the child's dash to the door can in no way be equated with the mad rush of a dog which has scented its quarry. A cautionary tale is that of a civil servant who one day decided to spread tons of napthalene pellets over an airport tarmac to get rid of the birds which were continually colliding with planes and posing a threat to them during take-off. The civil servant, with no knowledge of animal behaviour, was unaware that birds have no sense of smell. He did not realize that his drastic measure was pointless.

The sensory equipment of animals cannot be equated directly with our own. In some species, particular senses are far better developed than in humans, while other species have poorer development of certain sensory capacities. There are also animal species which have totally different senses. Bees, for example, can perceive ultraviolet light, and dogs can hear ultrasonic sounds. Humans can do neither of these things. In fact, most animals possess sense-organs of some kind which are lacking in human-beings, even if at first sight it seems that their visual, auditory or olfactory organs are far more "primitive" than our own.

The extraordinary sensory capacities of animals might quite easily lead us to think that if we were able to achieve a perfect understanding of the way in which they respond to stimuli (that is to say, those specific aspects of the environment which trigger their sense-organs), we would be able to understand all their behaviour. For a long time this was believed by some scientists who saw animals as "reflex machines". But this is not the case at all. According to the prevailing circumstances, an animal can easily perform a behaviour pattern in the absence of the stimulus which should "normally" evoke it. The irritation produced by a flea on a dog's belly elicits a scratching reflex involving rapid to-and-fro movements of one of the hindlimbs. Yet a slight touch on the same area of skin, perhaps when the dog is stroked by its master,

suffices to elicit this behaviour without any other stimulus. Similarly, even after the wings of a fly have been removed it will continue to make cleansing movements to remove dust particles from them. We must therefore accept that animal behaviour also depends upon internal factors, among which hormones are the most obvious. Indeed, the study of hormones has become a science in itself: endocrinology.

The Study of Animal Behaviour

In the years following the Second World War, i.e., since 1945, two quite distinct schools of thought have emerged, based on opposing approaches to the study of animal behaviour. The first school was founded in America by B. Watson and is known under the name of "behaviourism". It is essentially based on the results of laboratory research. Behaviourism is limited by its approach to the manipulation of behaviour, using standard experimental procedures and a very restricted range of animal species which are usually domesticated forms whose behaviour has already been modified by selective breeding.

The second school of thought on animal behaviour originated mainly in Europe. It was initially founded in the 1930s by the Austrian scientist Konrad Lorenz and subsequently reinforced by the work of the Dutch zoologist Niko Tinbergen. In 1973, these two shared with the German scientist Karl von Frisch the Nobel Prize in medicine and physiology. Among other things, von Frisch provided detailed explanations of the behaviour of bees.

After the war, Lorenz and Tinbergen founded a group of "field-workers" who conducted their observations of animal behaviour under natural conditions. These ethologists, aided by modern technology, such as radio-transmitters (attached to both terrestrial and aquatic mammals) and sound-recording equipment, have made numerous important discoveries.

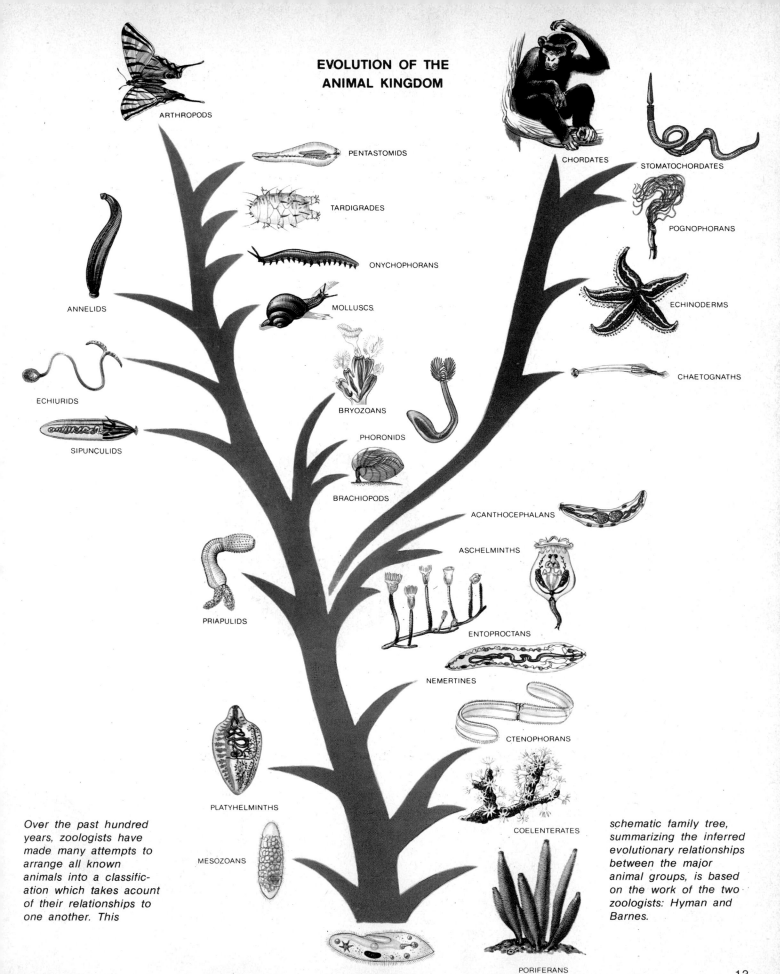

EVOLUTION OF THE ANIMAL KINGDOM

ARTHROPODS

PENTASTOMIDS

TARDIGRADES

ONYCHOPHORANS

MOLLUSCS

ANNELIDS

ECHIURIDS

BRYOZOANS

SIPUNCULIDS

PHORONIDS

BRACHIOPODS

CHORDATES

STOMATOCHORDATES

POGNOPHORANS

ECHINODERMS

CHAETOGNATHS

ACANTHOCEPHALANS

ASCHELMINTHS

PRIAPULIDS

ENTOPROCTANS

NEMERTINES

CTENOPHORANS

PLATYHELMINTHS

COELENTERATES

MESOZOANS

Over the past hundred years, zoologists have made many attempts to arrange all known animals into a classific- ation which takes acount of their relationships to one another. This

schematic family tree, summarizing the inferred evolutionary relationships between the major animal groups, is based on the work of the two zoologists: Hyman and Barnes.

PROTOZOANS

PORIFERANS

13

Protozoans (Single-celled animals)

This group contains the most primitive organisms found in the animal kingdom. It is composed entirely of single-celled organisms which often combine to form colonies. They exhibit a considerable variety of form and life-style. A tiny protozoan belonging to the category known as radiolarians (left) and (below) a slipper-animalcule (Paramecium), provide two examples of the diversity found in this group. The slipper-animalcule is covered with tiny, hair-like cilia which enable the animal to move around, whereas radiolarians have only rigid pseudopods (false feet).

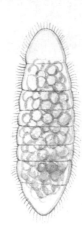

Mesozoans

This is a small group of parasitic marine organisms which differ from the protozoans in the incorporation of several cells into their bodies. One part of the body (the reproductive cells) can be renewed and perpetually produces germ cells while the rest of the body (the somatic cells) cannot be regenerated. These animals represent the transition between uni-cellular organisms and small multicellular animals.

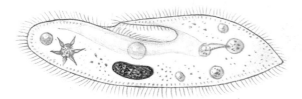

Platyhelminths (Flatworms)

This group includes a variety of so-called "flat-worms" found in sea-water, in fresh-water or in hot, humid environments. They sometimes occur as parasites. The tape-worm belongs to a special sub-group of the platyhelminths known as cestodes, which lack a digestive tract and absorb their food directly through the skin. The large liver-fluke (right), occurs as a parasite in both sheep and human-beings.

Poriferans (Sponges)

The poriferans, or sponges, are an essentially marine group of animals. They are multicellular animals with a considerable degree of different-iation between individual cells. They are character-ized by an internal cavity lined with special cells (known as choanocytes) and by the possession of a skeleton consisting of tiny spicules, which gives the body its rigidity. Approximately three thousand species can be found distributed through the seas and oceans of the world. The different species are classified into sub-groups according to the material constituting their skeleton, which may be reinforced by calcium, silica or horny material.

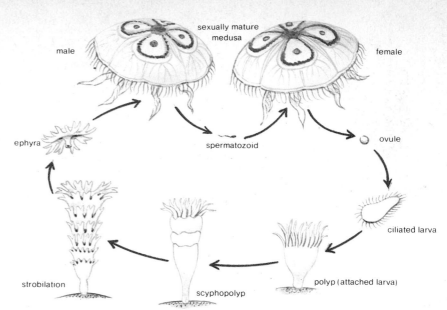

male · sexually mature medusa · female
ephyra
spermatozoid
ovule
ciliated larva
strobilation
scyphopolyp
polyp (attached larva)

Coelenterates

The body of these animals can be organized in two different ways, as a medusa or as a polyp. The reproductive cycle of the polyps is very complex. They are attached to the sea-bed or to rocks and they reproduce asexually, producing a colony of interconnected individuals. At the medusa stage, two animals of different sex are required for breeding, but they must pass through a "polyp" stage before becoming adult (top right).

Ctenophorans (Combjellies)

The animals constituting this group, which comprises some eighty marine species, are transparent and have a gelatinous appearance. They move around rapidly with the aid of lines of pulsating cilia which are arranged along the body margin. These animals represent an important stage in the evolution of the invertebrates. Above: a common Mediterranean species which may grow to more than one metre (three feet) in length.

Nemertines (Ribbonworms)

These animals are characterized by a well-developed digestive tract, including an elongated stomach and an intestine with a number of ramifications. Some nemertine species are parasites of bivalve molluscs, while others live in humid soil. Certain species may reach a length of one metre (three feet). When they are severely irritated, nemertines may fragment into a number of smaller pieces, each of which can regenerate to form a new individual if conditions are favourable.

Aschelminths

These are tiny animals which bear at their anterior end a circular organ fringed with cilia. The body is covered with a transparent membrane which in some species may be thickened to form a veritable carapace, known as a lorica. The animal illustrated here (right) is a rotifer which is less than two millimetres (one-twelfth of an inch) in diameter. Rotifers reproduce by a form of "virgin birth" (parthenogenesis) through which females can produce developing eggs without mating.

15

Entoproctans

The entoproctan (below) belongs to the pedicelline sub-group in which colonies are formed by a number of individuals, each consisting of a peduncle bearing a cup-like body containing the vital organs: tentacles, cerebral ganglion, mouth, stomach, intestine and anus. Both freshwater and marine species can be found. (The name "entoproctan" is derived from Greek roots signifying "enclosed anus".)

Phoronids (Horseshoe worms)

This is a small group of animals belonging to the general lophophorate category, represented by only a dozen species. They are tubular and wormlike in shape, with a U-shaped digestive tube and a double whorl of tentacles surrounding both the mouth and the anus. These animals are quite common in shallow sea areas, each individual living in a tube which it secretes for the purpose.

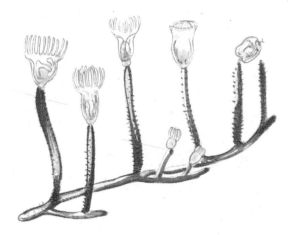

Acanthocephalans (Thorny-headed worms)

Acanthocephalans, which live as parasites in the digestive tracts of vertebrates, have a wormlike body which may be from six millimetres to sixty-five centimetres (one-quarter of an inch to twenty-six inches) in length. The anterior end is equipped with a crown of recurved hooklets which enable them to attach themselves to the walls of the intestine in which they obtain their food by simply absorbing it through their body surfaces.

Priapulids

These small cylindrical, wormlike animals live buried in the sand beneath cold seas. They are characterized by a rectractable proboscis at the anterior end and by a posterior forked "tail" of unknown function. The nervous system consists of a clearly demarcated ventral cord, but these small animals (only a few centimetres long) lack a circulatory system.

Brachiopods (Lampshells)

Each individual brachiopod is contained in a bivalve shell produced by a special tissue known as the "mantle". The body, which is partially enclosed in the shell includes an adhesive peduncle (left). The two valves, which are ornamented with fine grooves and indentations, are maintained in position by special muscles and some species do not even have a hinge between the valves.

Annelids (Ringed worms)

This branch of the animal kingdom contains marine forms, such as polychaete worms; terrestrial forms, such as the earthworm; and freshwater forms, such as the leech (right). All annelids share a common body pattern in which there is a series of ring-like segments (hence the name). In certain polychaete worm species, reproduction can involve strange dances which are performed in association with the lunar cycle.

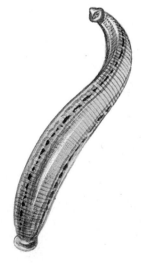

Bryozoans (Moss-animals)

Bryozoans, which can be found in fresh, brackish or salt water, cover pebbles with a kind of felt-like, soft moss and their name in fact means "moss-animals". The illustration (right) shows Plumatella, which can often be found in sheltered places in pools and water-courses.

Molluscs

These animals, which generally carry a well-developed external shell, show a tremendous range of body sizes from a few millimetres to the seventeen metres (fifty feet) recorded for some cephalopods. The dorsal surface of the body is covered by a solid cutaneous mantle, while the ventral surface is very muscular and is typically transformed into a "foot" used for digging, swimming or anchoring. From a common ancestor which must have lived on the rocky sea-floor, a number of separate evolutionary lines have developed, producing the six major sub-groups which are now recognized (right). Below: a typical gastropod mollusc, the snail.

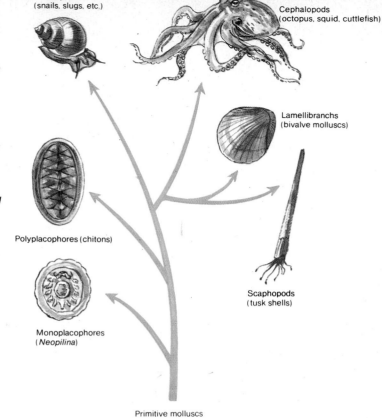

Gastropods
(snails, slugs, etc.)

Cephalopods
(octopus, squid, cuttlefish)

Lamellibranchs
(bivalve molluscs)

Polyplacophores (chitons)

Scaphopods
(tusk shells)

Monoplacophores
(*Neopilina*)

Primitive molluscs

Echiurids

Together with the tardigrades, the sipunculids, the onychophorans and the pentastomids, these animals constitute a side-branch of the main evolutionary stem leading to the molluscs, annelids and arthropods. These five groups together, which are sometimes referred to collectively as the pararthropods, mark the transition from the annelids to the arthropods and they are undoubtedly derived from a common ancestral stock.

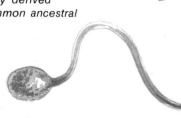

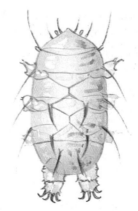

Onychophorans

These centipede-like terrestrial animals, which were first described in 1828, have a cylindrical segmented body with thirteen to fourteen pairs of legs. There are very few species in this group, a typical representative (below) being the "giant" Peripatus which grows up to ten centimetres (four inches long) and lives on the forest floor in Panama, feeding on small animal prey. Peripatus is only active at night, spending the daytime under leaves or buried in humus, so it is a very difficult animal to observe.

Tardigrades

Tardigrades, which are barely more than a millimetre (one-twentieth of an inch) in length, are able to survive extremes of dehydration and high or low temperature. They can support temperatures ranging from 100°C (the temperature of boiling water) down to −272°C (the temperature of liquid helium).

Sipunculids (Peanut worms)

The species in this group, which all possess a U-shaped digestive tract with considerable convolutions, are exclusively marine, living in sandy substrates and rock crevices down to depths of 4,000 metres (15,000 feet). Since they lack any kind of differentiated respiratory system, the digestive tract practically floats in their liquid-filled bodies. Sipunculids live on detritus and on tiny living organisms.

17

Echinoderms

These are among the best known of the marine invertebrates, comprising the starfish (below), the sea-cucumbers, the sea-urchins and the brittle-stars. More than five thousand species are known altogether. Despite their great diversity of form, all the echinoderms show a family resemblance in the radial symmetry which they have inherited from their common ancestors.

Arthropods

The arthropods, which include many thousands of different species, are currently the most successful animal group in terms of colonization of the Earth's surface. They are found both on land and in the air and have colonized both marine and freshwater environments. This group includes the insects (comprising more than two-thirds of the known animal species) and the crustaceans. The external skeleton of the arthropods, referred to as an exoskeleton, is a hard carapace which is often impregnated with mineral salts and with various pigments which determine the typical colour pattern.

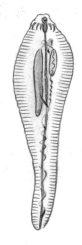

Chaetognaths (Arrow-worms)

The sea-living species belonging to this group never exceed ten centimetres (four inches) in length. Their bodies are divided into three regions: a head equipped with prehensile hooks, a trunk bearing two pairs of lateral, horizontal fins, and a tail which is also finned. This typical division of the body is already apparent during embryonic development.

Pentastomids

Almost all pentastomids live as parasites in the lungs and nasal cavities of carnivorous vertebrates. However there are also a number of species which parasitize large African and South American reptiles (including crocodiles), certain fish and even human-beings. The larvae pass through a number of distinct stages to produce adults of separate sexes, the males being smaller than the females.

Stomatochordates

These are all marine animals forming a small zoological group with only eighty species. The thread-like body is divided into three distinct regions. Respiration is carried out by means of gill pouches. This group includes the enteropneustans (below) — literally: "gut-breathers" — which possess a straight digestive tract and can grow to a length of 2.5 metres (eight feet). Their circulatory system incorporates a kind of heart in the form of a sac with contractile walls. Study of their embryos has shown that the stomatochordates are related to the echinoderms and pogonophorans.

Pogonophorans

These animals (left) are very closely related to the stomatochordates and are similarly exclusively marine in habits. They also have thread-like bodies and can grow up to 38 centimetres (15 inches) in length, despite being only 0.8 millimetres (one-thirtieth of an inch) in diameter. These animals do not possess a mouth or a digestive tract except during their larval development.

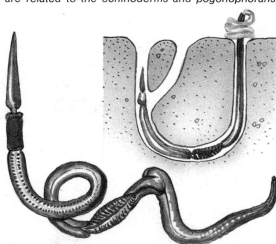

Chordates

This zoological group, which has undergone a remarkable evolutionary radiation, includes a number of unusual marine forms, such as the sea-squirts (ascidians), as well as the entire range of the vertebrates: fish (both bony and cartilaginous species), amphibians, reptiles, birds and mammals. All of the chordates have an internal skeletal element which supports the body and a tubular central nervous system. These two features are found in all chordate species, at least during their early development, even though in a few rare cases modification occurs in the adult stage. This is why this particular group of animals contains such a diversity of species, ranging from fish to human-beings. The chordates, which have been present among the Earth's fauna since the Palaeozoic era, now comprise more than 37,000 modern species which have occupied a vast array of different habitats.

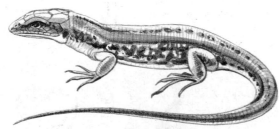

Instinct and Adaptation

According to a widespread popular view, human-beings are supposed to act under the control of their intelligence while animals are governed by instinct. In fact, animals in general possess at birth the majority of the behaviour patterns which they will follow in response to specific stimuli, whereas human-beings acquire most of their behavioural characteristics after birth. However, many animals do not retain the same patterns of behaviour unchanged throughout their lives. In other words, in addition to its initial complement of innate behaviour patterns, an animal will generally acquire additional behavioural features in the course of its lifetime. Animals can "learn" to respond in a particular way to stimuli which did not previously evoke a response.

Programmed Behaviour

In fact, animal behaviour is programmed by a combination of internal and external factors. Internal programming progresses predictably with an animal's development, while external programming depends on each individual's particular experience of the environment. Both kinds of programming result from a process of trial-and-error, but internal programming actually takes account of the "lessons" which the species as a whole has learned in the course of its evolution through successive generations. External programming, on the other hand, depends upon the individual's experience in the course of its

own lifetime not on that of its ancestors. Thus it is that a young chick will initially respond "instinctively" by crouching if a flying object passes overhead, regardless of whether this is a dead leaf, a piece of paper carried away by the wind, a small bird, or a bird of prey. As the chick grows up, it will crouch less and less. Is this because it has simply become "habituated"? Not at all. The chick has simply learned that no danger is associated with a falling leaf or with a commonly seen swallow. By contrast, the chick will continue to respond predictably to a bird of prey, not because it is "instinctively" aware of the danger, but because it considers as dangerous any unfamiliar "flying object".

The survival of many small animals which must defend themselves against predators depends upon this ability to learn from experience. Many insects exhibit what the scientist calls "warning colouration", a signalling that they are unpalatable, that they are poisonous, or simply that they have a very hard carapace. All birds learn not to touch prey with warning colouration, but only after trying to eat one or two. They eventually learn to modify their behaviour in response to the original stimulus ("insect = edible prey").

Although it is relatively rare in animals, there are cases of learning through imitation. For instance, if chaffinches are reared in isolation from adult males of their own species they will produce completely disorganized

In order to remove external parasites, birds have dust-baths or expose themselves to the sun. The jay makes use of ants in cleaning its feathers, holding the insects in its beak so that their formic acid will attack the parasites. Birds make sure that their feathers remain waterproof by coating them with the secretions of their preen gland, located on the rump. Above: a darter (snake-bird) drying out its feathers.

Large mammals such as the zebra (below left) and the rhinoceros (below) are also plagued by a variety of parasites. Zebras roll on the ground in order to dislodge external parasites, while rhinoceroses take ponderous mud-baths so that a dry crust will form on their hide and provide a temporary shield against biting insects.

The big mammals of Africa visit water-holes to quench their thirst at the coolest times of the day, around dawn and dusk. The two big cats shown here drinking are the cheetah (right) and the lion (below).

vocalizations. It is only through hearing the calls of the adult male that they will learn to produce the species-specific song. There are also some bird species which learn the songs of other species, such as the famous blackbirds of Lyons which, up until about twenty years ago, imitated the whistles used by street-traders to advertise their wares. Niko Tinbergen tells the story of one of his German colleagues who had reared a young male bullfinch close to some cages containing male canaries. The bullfinch learned to imitate the canaries so well that it was impossible to distinguish his calls from those of a real canary. Later on, the bullfinch was mated with a female of his own species. A clutch was reared and two male fledglings also learned from their father to sing like canaries. The sequel to this story is even more extraordinary: one of the young males, which had been sent to a bird-fancier three kilometres (two miles) away was mated with a female bullfinch. When, two years later, one of the offspring from this pair was sent back to the place where his grandfather was reared, the owner was surprised to find that the young male was also able to sing exactly like a canary.

In addition to external stimuli, other stimuli

The techniques which animals use to obtain their food are extremely varied. In the African ibises (above) the tip of the bill curves downward slightly and this proves to be a very effective device for extracting bivalve molluscs (their main food source) from their shells. They insert the tip of the beak between the shells and remove the mollusc in a few seconds without breaking the shells. The black kite (left) inspects its territory by flying in large circles. As soon as a suitable prey is spotted, the kite plunges down to catch it. The male walrus (below left) uses his huge tusks to probe into the mud of the sea-bed and to dig out the small animals on which he feeds. The marmot (below) is a rodent whose diet consists entirely of grasses and roots.

are generated within an animal's body. This is why hormones are secreted by the pituitary gland, and the sex organs are the real internal agents governing reproductive behaviour. Nevertheless, the secretions of such glands are, to some extent at least, influenced by external factors such as variation in the length of the days during the year. In many vertebrate species, the glands involved in reproduction only become active in Spring. In fact, by habituating such animals to an artificial light regime in which "day-length" is gradually increased, it is possible to stimulate their pituitary glands and their sex organs to produce hormones even in winter. Throughout the entire reproductive process hormone secretions and external stimuli exhibit continual interplay and their influence is such that the appropriate behaviour patterns emerge

Insects show a great diversity of feeding habits, which are highly adapted to specific environmental conditions. There are many species which feed essentially on plants, like the grasshopper (centre right) and greenfly (right). Some insect species, including many butterflies and moths (top right) do not feed as adults, whereas their caterpillars may be veritable pests for farmers and foresters.

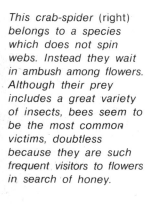

This crab-spider (right) belongs to a species which does not spin webs. Instead they wait in ambush among flowers. Although their prey includes a great variety of insects, bees seem to be the most common victims, doubtless because they are such frequent visitors to flowers in search of honey.

at the right times.

As with the specific case of reproduction, a great deal of animal behaviour is cyclical in form, consisting of a series of related actions. The search for food is no exception to this. The first stage undoubtedly involves an internal change which evokes a "spontaneous" action: the hungry animal begins to seek out food without requiring some scent to spark off the search. Thereafter, subsequent behaviour patterns are followed in sequence in response to specific stimuli. A hungry digger-wasp will first of all fly to an area where the usual prey (e.g. bees) can be found. The wasp flies from flower to flower, hunting "by sight" until a potential victim is spotted. The prey is then followed from downwind, still by visual means, until the wasp can approach close enough to identify the bee as such by smell. Only when this has happened will the digger-

wasp pounce on the bee and immobilize it with its sting. The entire hunting sequence takes place according to a sequence of stimuli which induce the wasp to pass from one behaviour pattern to the next. Initially, it is the movement of the potential victim which evokes the visual hunting response, and even a crude dummy will elicit the same behaviour. Thereafter, olfactory identification is involved as the digger-wasp approaches its prey. A dummy perfumed with the smell of a bee will elicit the final approach, while a digger-wasp will actually turn away from a bee which has been deodorized. It should be noted that the wasp responds to different kinds of stimuli according to the stage the hunt has reached; a suitable dummy must have not only the shape of a bee but also its odour.

Strangely, the successive activities in such a chain of response will often be performed

The behaviour of insects is essentially innate. That is to say, it is very largely inherited. The house-fly (above) which can often be seen vigorously rubbing its wings to remove dirt particles from them, will continue to exhibit this behaviour even when the wings have been removed.

25

The search for food and the avoidance of predators together provide much of the motivation for animal behaviour. Right: a hedgehog, raising its spines when confronted with an adder. Below: a nudibranch mollusc (seaslug) slowly roving over a coral in search of food. Facing page: a coyote feeding on the carcase of an American bison.

without any external control. Some male butterflies carry out a complex ceremony once they have found a female and the process culminates in mating once the scent organs of the male have come into contact with that of the female. The ceremony is performed in exactly the same way with a dead female, and even if the female is completely removed the male will continue, stopping only at the point where olfactory recognition would normally elicit the act of mating.

All the different sense organs which animals possess are brought into play in the search for food. Sight, hearing, smell and touch (involving contact with whiskers or tactile hairs) combine in various ways in different species. For a specialist in animal behaviour, the study of feeding activities is a difficult undertaking since the daily diet of most animal species is greatly influenced by the abundance or scarcity of available foods. In addition, it is not always easy to determine what a particular species is eating, particularly with small animals that are essentially nocturnal in habits. It could be thought that it would be possible to overcome these difficulties by studying the feeding behaviour of animals kept in cages or in large enclosures. As with numerous other aspects of behaviour, however, results obtained under such conditions might lead to serious misinterpretation. Some interesting observations have been made in this context by Professor Hediger of Berne Zoo in Switzerland, where the species on display to the public include not only indigenous animals such as chamois and ibex, but also African ruminant species. Hediger noticed that the latter often poisoned themselves by feeding on a weed which is common in western Europe, the deadly nightshade. Chamois and

Many animal species live in social groups of different sizes in order to co-operate in tackling the problems they face. Lions (above) hunt in groups and rear their young communally.

ibex, by contrast, carefully avoided the plant. The African ruminants had not encountered the deadly nightshade in their countries of origin and had to "learn" its properties. From this, it might be concluded that the diets which animals adopt are "learned" and do not involve "instinct" at all. However, if this were true, how could we explain the fact that a beaver from Scandinavia which stores food near its den every autumn to tide it over the winter will continue to behave in the same way when moved to more southerly latitudes? Yet, in the opposite situation a beaver originating from the Camargue region of southern France, where it does not store food for the winter, will begin to do so if moved to a cold northern region. In this case, as in others, it is very difficult to distinguish what is innate and what is learned. In any event, it is certain that the two types of behaviour, innate and learned, are combined together. The question then remains: how can an animal select the appropriate course of action when there is a divergence between what instinct dictates and what is learned from experience?

A Situation of Conflict

It would doubtless be impossible for an animal to perform a mixed behaviour pattern, given a conflict between different kinds of stimuli. An animal must select a response before taking action. A single course of action must be followed, or nothing would be achieved. Yet, in many instances, an animal must be subjected to several different stimuli simultaneously. In such cases, it can happen that none of the expected two or three alternative behavioural responses will occur. The animal finds itself in a "conflict situation" and this leads to what experts call "displacement activities" (substitute responses). A number of examples will illustrate differing conditions under which this can occur.

A young cock placed in a chicken coop comes face to face with an old hen, the established ruler. In response to the hen's aggressive pose, the cock flattens his feathers rather than acting more logically by either fighting or

fleeing. This is one example of "displacement behaviour". In other cases, the same cock in a conflict situation might show incomplete performance of a number of different behaviour patterns, such as rearing up or lowering its head as if preparing for combat, but then remaining in a fixed pose. Or, to take another example, the cock may be seen vigorously shaking a feather picked up from the ground, more-or-less diverting its feelings to a substitute object — another form of "displacement activity". In fact, this kind of activity is not restricted to animals. When we are hesitant about expressing our feelings, we may scratch our heads, our noses or our ear lobes. When we are angry, we may clench our fists or stamp our feet, but not actually come to blows. Everyone has seen the situation where a child, after being scolded, will tear up a piece of paper, or perhaps even a school notebook, to "work off its anger".

As a rule, conflict situations are short-lived under natural conditions. In the laboratory, however, it is possible to create situations which persist for some time so that behav-

The praying mantis (top left), a veritable predator, remains immobile and waits for its prey. Any insects which alight within range are seized between the anterior limbs, which are covered with spines. Centre: a snake in the process of swallowing an enormous frog, thanks to the mobile articulation of its jaw joints. Below: an otter, a mammal which has become a specialist in the capture of fish in the course of its remarkable adaptation for aquatic life.

In the course of their long evolutionary history, the primates (prosimians, monkeys and apes) have become beautifully adapted for life in the trees. The sifakas (above) which belong to the lemur group of Madagascar, sleep in trees at night, huddled together if it is cold. The orang-utan (below left) is an ape species which is now threatened by forest destruction in south-east Asia.

ioural consequences can be studied. Such research is fascinating, for all of us like to be able to find out how and why things happen. In addition, observations carried out on animals in this way are of great interest to the general public because they enable us to understand aspects of our own behaviour which are at first totally inexplicable. Man, too, has unexpected and bizarre forms of behaviour. Finding answers for such behavioural peculiarities can sometimes assist us in treating our fellow human-beings who are suffering from stress or depression, or in interpreting problems of delinquency which originate in "displacement behaviour".

In fact, systematic observation combined with appropriate experimentation permits us to understand how the behaviour of each species constitutes an integral part of its "survival mechanisms", which are essentially concerned with three factors: obtaining food, avoidance

The polar bear (above) would probably find it quite difficult to survive outside its normal habitat. Thanks to its white colour, which enables it to blend in with its environment, it is able to approach its prey unnoticed. Its thick fur and the layer of fat beneath the skin provide efficient protection against the cold.

Below: a colony of Adélie penguins, birds which have become so well adapted for aquatic life that their wings have been transformed into flippers and can no longer be used for flight.

In order to protect themselves against predators, many animals have developed special defensive weapons. The bumble-bee (right) has, in the course of its evolution, developed a robust sting. The South American tree-frog (below) possesses special skin glands which secrete an extremely powerful poison. There are numerous insects which have come to resemble stinging bees and wasps. Below centre: a fly which lives among flowers and looks just like a bee. Some caterpillars which are covered with irritating hairs (below right) are almost entirely avoided by birds thanks to this special protective shield.

of predation and successful reproduction.

For example, to what extent might we be able to modify aggressiveness in human-beings, and what educational procedures would be required? Indeed, just how appropriate are the methods of education and re-education that are currently employed? As yet, we can only provide a small part of the answers to these questions, but in the search for relevant information it is particularly valuable to study the behaviour of our closest zoological relatives, the great apes. We know that their behaviour is profoundly influenced by the relationship between mothers and their infants, as is the case with numerous other species. But this is also true of human-beings.

It is now known that, in order for an adult to show normal social behaviour, it is indispensable to have prior experience of maternal love, that is to say of the various patterns of care which an attentive mother shows towards her child. Such application of the findings of specialists of animal behaviour to the interpretation of human actions provides the most compelling justification for studies which are extremely time-consuming and which may not at first sight appear to be of great use. Although it is practically impossible to prove anything with the fossils that we have at our disposal, it is certain that animal behaviour has evolved hand-in-hand with other adaptations which permit animals to survive in

their natural habitats. The species which we see today owe their survival to the combined transformation of behaviour and structure during their evolution.

The appearance of the scorpion (below) is in itself enough to discourage a potential aggressor; its possession of a poisonous sting adds the finishing touch. Coral reef fish (right), which seem to be extremely brightly-coloured, are in fact less obvious against the luxuriance of their natural background. In a number of cases, however, the striking colour patterns act as a warning to predators that the fish as a whole is poisonous or that it bears poisonous spines.

Territory and Aggression

Most solitary terrestrial animals show attachment to specific areas where they have access to seasonal or permanent resources. They depend upon habitual use of these areas to feed, to shelter, to sleep and to breed. In many cases, such a solitary animal will show great intolerance to other members of the same species, and will vigorously defend the home area against any encroachment. Such aggressiveness is the most typical manifestation of territorial behaviour, though from a human point of view it may seem to be simply anti-social. In fact, territoriality plays an extremely important part in the survival of many species.

"Controlled" Aggression

It emerges in many cases that such territorial aggression, which has become accentuated during the breeding season, is considerably less deadly than one might imagine. In fact, it is quite rare that jousting between two stags during the rut leads to the death of one of the combatants, or even to serious wounds. Everything takes place as if the aggression were being kept under tight rein. However, this kind of control should not be confused with that shown by two nations which reach the brink of war and then decide against it for economic or humanitarian reasons. We should not forget that animals are incapable of rational thought. It is significant that such "control" of animal aggression is most noticeable among animals living in families, groups or colonies which would be rapidly disrupted by persistent aggressiveness among their members. So such aggression, while it continues to exist, has generally been transformed

For many animal species the territory must be extremely important for survival, since a great deal of time and energy is spent in defending it. During the breeding season, the males of such species (e.g. deer and their relatives) will confront other males on veritable duelling grounds. Territorial defence behaviour is particularly common among mammals, birds, reptiles and fish. Left: a giraffe. Right: a female white-tailed deer — the North American equivalent of the roe-deer.

into a kind of ceremony — it has become "ritualized".

While most fighting among animals, be it sham or real, occurs during the breeding season (when male hormones increase the level of aggressiveness), combats also take place within animal social groups in the struggle over social rank in the hierarchy.

A wide variety of natural weapons may be used by animals involved in such fighting behaviour. Among mammals, big herbivores have horns, antlers or heavy hooves, while carnivores have powerful canine teeth and sometimes claws as well. In reptiles and fish, teeth can also play an important part and the tail can be used to batter the opponent. The hook which develops on the lower jaw of the male salmon during the spawning season similarly acts as an offensive weapon. Other fish thresh their tails in the water, producing turbulence which will adversely affect the sensitive lateral line system of an opponent.

Such fighting during the breeding season rarely has fatal consequences. In many cases, interactions are confined to intimidatory displays which can take a wide variety of forms. They may be limited to particular postures signifying threat or they may involve special vocalizations. If carried out on the territorial boundary, these displays can act as a form of territorial marking, just like the deposits of urine or faeces made by foxes and wolves, or like the special glandular secretions of some species; for example, the castoreum of the beaver. When the fights which take place during the breeding season are analyzed, it emerges that they always take place between distinct categories of individuals. Adult males and adult females, for instance, engage in quite separate confrontations. Territorial encounters involving threat gestures ensure that the territory, which will only supply a limited amount of food and shelter, will be occupied by only one individual. These encounters generally take place in a limited area, which may be close to the nest-site in certain bird species or in the presence of females, as can be the case with members of the deer family (Cervidae).

The term territory should not be confused with the concept of "home range", which is simply the area regularly used by an animal or by a social group of animals. The home range differs from a territory in that it is not necessarily defended against encroachment by other members of the same species.

The Nature of Aggressiveness

Psychologists and other specialists in animal behaviour have long been interested in the nature of aggressiveness. By studying this in animals, we can carry out comparisons with human aggression, which plays a large part in our social interactions. It is virtually certain, on the basis of evidence already collected, that aggression has some kind of hereditary basis, but it is equally certain that it can be canalized — if not completely suppressed — by appropriate training or education. Recent studies have contributed greatly to our understanding of aggressive behaviour, as has been set out in Konrad Lorenz's book devoted to this subject.

The three-spined stickleback, commonly found in ponds, pools and streams is a very familiar sight. It has now been intensively studied in the laboratory and has provided many insights into the control of animal behaviour. Studies of the stickleback have revealed the importance of what scientists refer to as "sign-stimuli", which are special signals releasing specific responses from the male and female during the breeding season. For most of the year, the male is greyish-green with darker bands over the body. Gradual increase in day-length in Spring activates the production of reproductive hormones which bring about red pigmentation of the belly (right). This special colouration acts as a prenuptial signal for the female and as an intimidatory signal for male rivals competing for the territory. The red colour of the body is then accompanied by a blue colouration around the eye and a bluish hue of the dorsal surface, which acts as a signal attracting a female into the male's territory. The male only regains his original dull colouration after the eggs have hatched, when his inconspicuous colour will protect him as he looks after the offspring. The male remains with the young after hatching and will, for example, bring them back in his mouth if they start to wander off.

Although the exact functions of territorial behaviour are not yet fully understood, it is likely that it ensures that animals are evenly distributed throughout the species range and that each individual will therefore have access to the required food resources. Nevertheless, territorial behaviour is not necessarily connected with the presence of food. Although this is so, for example, in the case of territorial demarcation among birds of prey, there are many sea-bird species, such as cormorants, which will savagely defend a patch of rocky terrain which is completely devoid of food resources. In such cases, territorial defence presumably functions to limit the number of individuals living in a colony and feeding in the nearby coastal waters. It also prevents overcrowding of the nests themselves.

Frogs and toads (batrachians) comprise one of the few animal groups in which territorial behaviour is very rare. The tree-frog (below) uses calls for communication rather than territorial defence. Elephants (right) become particularly aggressive during the breeding season.

According to Lorenz, aggressiveness in both man and animals is the result of innate tendency, that is, based on genetic inheritance. The characteristics of aggression are comparable to the physiological motivation which drives animals and human-beings to eat and drink. Some research workers, while accepting the existence of an hereditary basis of some kind, suggest that human aggressiveness is dependent upon individual experience, that is to say that external factors are more important than internal ones.

When the behaviour of different animal species is compared, it is found that it is not only solitary animals, living under "dispersed" conditions that keep their distance from one another. Social animals also maintain a minimum distance, thus setting a certain limit to their communal life. Seagulls, crows and starlings, which are all social species, often squabble with one another over food.

Except in a few cases where they result in a

draw, territorial combats end in victory for one of the opponents, for whom such behaviour pays obvious dividends. But there must be a loser, and there could be some advantage in losing, too. In fact, there is an advantage in knowing when to give up a fight as well as knowing when to attack. A male who withdraws from an unequal fight has more chance of producing descendants by moving onto an unoccupied area than one who wastes time and energy engaging in a hopeless fight with a superior opponent. Survival depends more on a balance between attacking and fleeing than on systematic, unrestricted aggression.

The Father of Modern Ethology

Professor Konrad Lorenz, awarded the Nobel Prize in physiology and medicine in 1973, is universally recognized for his studies of animal behaviour, in which he has established himself as one of the most brilliant specialists of our time. Until recently Director of the Max Planck Institute for the study of animal behaviour in Seewiesen (Bavaria) and Professor at the University of Munich, he has written numerous books including a veritable natural history of conflict: *On Aggression*. The science of animal behaviour was, at the outset, divided into two distinct schools. The European school of ethologists devoted itself to the study of instinctive behaviour on the basis of observations conducted under natural conditions. The American school of behaviourists or animal psychologists was primarily concerned with laboratory studies. Both schools have paid homage to Konrad Lorenz, who has been referred to by Julian Huxley as the "father of modern ethology". The basic lines of research which Konrad Lorenz established are still continued today. The new school of ethology which he championed is based on the conviction that an animal's behaviour, just like its physical adaptations, is an integral part of its survival equipment and has been subject to the same process of evolutionary adaptation. Demonstration of this fact rests upon a vast wealth of observations conducted on a great variety of animal species. In each case, the observer has been obliged to share the animal's daily life, including learning its "language" in order to gain a proper understanding. Work of this kind has been conducted, for example, on frogs, geese, ducks, shrews, dogs, cats and monkeys, always under the appropriate natural conditions.

An animal may also be aggressive towards its own offspring. One might at first sight believe that this would run counter to the natural law according to which the parents, or at least one of them (either the father, as in the case of the stickleback, or the mother, as in the case of mammals generally), provide care and protection for the young. Yet this is particularly evident in family-living species with a very close-knit social life, such as beavers. With these mammals, which exhibit very marked territorial behaviour and respond to any intrusion by other members of the species with vigorous attacks, the young are forcefully driven away once they have entered the third year of life. They are driven off so forcefully, in fact, that any young which resist receive vicious bites. The explanation for this is that they are no longer "young", but have become sexually mature beavers which, as such, are rivals to their own parents, competing with them for resources. It often happens that, if the adult male or female disappears at this particular time, the survivor from the original pair will establish a new family with one of the "young".

The aggression shown by parents to any stranger which attempts to approach their offspring has also been the subject of detailed study. Such aggression can sometimes be extremely violent, even when the parent is quite small and is attacking a member of a larger species. The famous animal photographer Eric Hosking had this brought home to him with tragic results on one occasion when trying to photograph some young birds of prey in their nest. The mother attacked him so fiercely that she blinded him in one eye. Even the female of the relatively small eared owl will not hesitate to attack a human-being to protect her clutch.

In one of his books, Konrad Lorenz tells a

strange story which at first sight one might be tempted to describe in an anthropomorphic way as a case of animal savagery. A female shelduck walking along with her ducklings spotted its owner holding in his hands a tiny, freshly-hatched mallard duckling which was uttering distress calls. Leaving her own ducklings, she rushed towards the man and attacked him so courageously that he was forced to drop the mallard hatchling. The sequel to the story is even stranger. The young mallard mingled with the shelduck's offspring and the female then attacked him so fiercely that he would have been killed on the spot if the owner had not intervened in time. Naturally, those watching this sequence of events, who had been admiring the female's initial response, were quite perplexed. What had happened? "The explanation of such apparently contradictory behaviour is quite simple," writes Lorenz. "The distress calls of the mallard hatchling are almost identical to those of a young sheldrake and the defensive response of the female is therefore stimulated automatically. But the plumage of a young mallard is quite different from that of a sheldrake, so it is identified as a stranger and this evokes another automatic response, that of protecting the clutch against an intruder. Thus, the young mallard which at one point was a lost infant in need of help an instant later became an intruder to be chased away.

The Gorilla, a Gentle Giant

Having examined the broad outlines of territorial and aggressive behaviour, we can now turn to some instructive individual cases.

When we speak of aggressiveness in animals, the species which is most feared and reviled (or at least used to be) is certainly the gorilla. The reports brought back by explorers and hunters during the last century all agreed in describing the gorilla as a creature of proverbial ferocity and prodigious strength, as an animal quite incapable of controlling its aggressiveness.

Over the past twenty years or so, scientists have come to reject this universal view of the great anthropoid ape, one of the closest relatives of man on the evolutionary scale. This reversal of opinion is particulary due to George B. Schaller, now attached to the New York Zoological Society. Thanks to him, the myth of the "monstrous gorilla" has been relegated to history. But, in order to achieve this, Schaller and his colleagues had to spend an entire year with the gorillas — a surprising adventure which he has described in a number of books. More recently, another American naturalist, Dian Fossey, has conducted an even more extensive study of the mountain gorilla, spending five years observing the species while living in a hut she had built in the forest. After a number of difficult and fruitless attempts to approach the gorillas this young woman succeeded in winning the confidence of the big animals by imitating their gestures and their sounds so that she became accepted as one of them. The observations conducted by George Schaller and Dian Fossey are packed with interesting facts. For example, the home range of a group of mountain gorillas, which will normally contain between five and twenty individuals, covers an area of about thirteen square kilometres (five square miles). Lowland gorilla groups contain a greater number of animals, between fifteen and thirty, and range over an area of twenty-four to forty square kilometres (nine to sixteen square miles). Gorillas move round over long-established pathways and each group is led by an adult male whose back is covered with white hairs which only appear when full maturity is reached, thus earning the name "silverback".

The adult males and younger members of the group keep a lookout and provide defence when required so that the females and their young benefit from continuous protection. Except when they are uneasy for some reason, the members of the group laze around or may occasionally play, sliding down trunks and mounds or chasing one another.

The territorial behaviour of gorillas is particularly interesting. Observers have noted that these anthropoid apes never fight with one another over food or in order to maintain exclusive rights over a particular area of land.

The crab (right) is displaying a typical threat posture. The Indian Ocean goby (below left) engages in fights using its teeth and flailing the water with its tail. Two hippopotamuses (below right) often become locked in battle without leaving the river where they spend most of their time.

Lead males of individual gorilla groups hardly ever engage in fights. Whenever they meet, two such dominant males will eye one another in silence with a threatening stare. In order to demonstrate its pacific intentions, a gorilla which encounters another will look towards it and nod its head. If one of the two gorillas in such an encounter should adopt a threatening posture, the other will usually move on or adopt a submissive posture. The threat behaviour of the gorilla provides a good example of the ritualization of aggression. Nevertheless, these threats can be quite spectacular and it is perhaps understandable that they were misinterpreted by early explorers and hunters. In the full threat sequence, a gorilla will rear up on its hindlegs, drum its cupped hands on its chest and then charge through the forest, uttering high-pitched screams and breaking branches or ripping up trees on its way. Given the size and strength of a gorilla, this exhibition is obviously quite terrifying. The

sudden attacks inflicted on early travellers by gorillas were a result of complete misunderstanding of the natural behaviour of these great apes, which only become dangerous when they feel threatened.

Much of the fighting behaviour that animals exhibit takes place during the breeding season. Males usually confront one another in a highly ritualized fashion and fighting rarely leads to the death of one of the combatants Some fighting also occurs in connection with food, as is the case among carnivores particularly. Two gnus (above) engage in a fight by using their horns, while brown bears (below left) and lions (below right) display their canine teeth as a sign of aggressiveness when involved in a dispute over prey.

Reproductive Behaviour

Animal reproduction is controlled by a number of different factors, notably by gradual changes in day-length over the year and by internal changes governed by hormones.

The Functions of Reproductive Behaviour

The basic function of reproductive behaviour is to ensure the continued survival of the species. Observations of reproductive behaviour show an enormous range of different adaptations. With many animal species, notably among the fauna inhabiting the seas, we can in fact scarcely talk of reproductive behaviour in view of the manner in which fertilization takes place. Take, for example, the case of the oyster. According to the ambient temperature and nutrient availability in its biotope, an oyster changes sex throughout its life. At the age of eight to ten months, given a summer temperature of between 15°C and 16°C, the oyster will be female and will become male only intermittently. If the summer temperature is between 20°C and 22°C, the oyster will change sex every year. Fertilization takes place in summer and the eggs, numbering about a million, are actually fertilized in the female's gill cavity, where they will remain until they hatch out. In this case, we cannot really talk of reproductive behaviour.

In most animal species, particularly those that live on land, fertilization depends upon a proper act of mating, or copulation, between two individuals of different sex. The act of mating involves physical contact and therefore runs against the grain of habitual, instinctive defence or fleeing. Such responses must obviously be suppressed for mating to occur at all. This explains why so many animals display a certain amount of aggression during

Animals will generally mate year after year, propagating their own kind without any awareness of the implications of their actions. The male lion (above) is following a female in heat, but she is not yet fully receptive. Even when the act of mating takes place (below), it is accompanied by a certain degree of aggressiveness.

During the breeding season of the elephant-seal (below), males are the first to move onto dry land to establish territories which they fiercely defend. The females join them a month later. Only those males which have been able to locate and maintain a territory will be able to receive and mate with a group of females. Red foxes (right) engage in a nuptial ceremony before mating.

the breeding season, even towards the partner with whom copulation will eventually take place.

Such aggressiveness seems to be increased even further when there are encounters between species of similar appearance; for instance, two bird species with almost identical plumage. This is also important for a species, since the elimination of any risk of cross-breeding between different species will prevent the occurrence of infertile hybrids. It is easy to understand why each species has developed an array of special peculiarities so that individuals can recognize one another reliably during the breeding season. Such special features include nuptial colour patterns, dances and peculiar vocalizations.

Certain tropical tree-frog species are almost identical in external appearance and the only barrier to breeding between the species resides in the males' calls. But much remains to be discovered about the sexual behaviour of animals, even if scientists have been able to establish that — with the exception of man

A complex ceremony of nuptial parading and intricate dances precedes mating in the wood grouse (above). Sometimes, a male becomes so aroused sexually that he will parade along roads and even attack human-beings.

and the great apes — the sex drive is dependent upon a combination of internal stimuli of hormonal origin and external stimuli such as changes in day-length.

Sunlight, temperature and rainfall all combine with internal changes in the organism to determine the timing of the breeding season. The advent of warm weather or of the rains can vary from year to year and the amount of available light can vary according to the cloud cover. The control centre of the vertebrate central nervous system, a special region of the brain known as the hypothalamus, responds to the gradual increase in day-length during Spring and transmits this information to the pituitary, a small gland located on the ventral surface of the brain. Subsequently, under the influence of various hormones, a series of changes take place which prepare the animal's body for reproduction.

Stimulating Mating Behaviour

As has been seen, mating can only take place under certain conditions. The two partners must first be drawn into proximity by means of stimuli acting at a distance, as with the visual signals of the stickleback. Subsequently, the sexual activity of the male and female must be synchronized so that the partners are "ready" at the same time. Finally, it is essential that both partners should cease to exhibit their usual aggressive or fear responses.

The increased aggressiveness exhibited during the breeding season between animals of different species considerably reduces the risk of hybridization. In addition, each species has specific "displays" which permit recognition of members of the same species by providing unequivocal information about the identity of the displaying animal.

In bird species, the female is commonly attracted to the male's territory by his special calls. In butterflies, on the other hand, there are extremely efficient scent signals which bring about the necessary encounters between male and female. Some male spiders produce specific vibrations on the silk threads of the web so that the female can recognize them, though this does not always prevent them from being eaten! Visual signals, such as the red belly of the male stickleback, also play an important part, and there are a number of special cases such as the light signals produced by glow-worms and fireflies.

Visual signals can be enhanced by particular patterns of movement (dances, leaping, special arrays of plumage, etc.). This is true of the peacock when it fans out its feathers and of the robin when it puffs out its chest in order to emphasise the red colour of the breast feathers.

In most cases, however, it is not a single signal which ensures recognition between the partners prior to mating, but a co-ordinated series of displays and responses exchanged between them.

Types of Mating Behaviour

The emission of signals is not always confined to the male. Strangely, the roles are actually reversed in a number of cases. In the fighting quail, for instance, it is the female which performs a parade and calls to attract the males. In fact, she mates with several different partners. This is a rare case of inverted sexual

The North American male sage-grouse (below) fluff up their plumage and thus increase their apparent size when defending their territories.

behaviour which is also found in a few other species such as the red-necked phalarope, a small wader which lives in Arctic areas and flies over Europe in autumn when migrating to Africa. In this species, the female has a nuptial plumage which is more colourful than that of the male. It is the female which defends the territory and attracts the male for mating.

Pair-formation can take place in a number of different ways. Konrad Lorenz recognizes three different categories: in lizards, the male parades his nuptial colours in front of any members of the same species. Females take flight and are then followed by the male. In some fish species, males and females together participate in spectacular demonstrations in the course of which a veritable hierarchy is established among the males. Finally, there are other species such as the angel-fish, which is well-known to aquarium-owners, in which the male and female threaten one another incessantly (sometimes with fatal consequences) in order to elicit the specific signals for mating.

Types of Fertilization

Fertilization requires an encounter between male and female gametes (between sperm-atozoa and ova). These gametes can meet outside the bodies of the animals concerned, giving rise to external fertilization. Naturally, this can only take place in water. The gametes can also meet within the female's body. This is known as internal fertilization and involves the release of spermatozoa by the male within the female reproductive tract, either as sperm or as spermatophores (small sac-like structures containing spermatozoa). However strange it may seem, both kinds of insemination can occur with or without actual copulation, so that four different types of fertilization can be recognized:

1. *External Fertilization without Copulation*
This is practised by coelenterates, echino-derms, aquatic molluscs, sea-living annelid worms and fish. Both the sperm and the eggs are liberated into the water for fertilization to take place.

2. External Fertilization with Copulation

This can be observed with frogs and toads. The male will clasp the female, often for several days, and release his sperm onto the eggs as they are laid. Certain South American tree-frogs which mate out of the water roll a leaf into a kind of container and the sperm and the eggs are deposited inside.

3. Direct Internal Fertilization

This is the most frequent form of fertilization found and it is the standard pattern for mammals. The male's copulatory organ (penis) is introduced into the genital orifice of the female (vulva) and the spermatozoa are deposited directly inside the female tract (vagina). In the case of birds, with a very few exceptions, the male and female simply juxtapose their cloacas.

4. Indirect Internal Fertilization

This last type of fertilization is usually only possible in a humid environment. The male deposits his spermatozoa outside the female's body and they find their way into her reproductive tract themselves in order to fertilize the eggs. This form of behaviour is found with newts in particular. The male releases into the water a spermatophore which sticks to the female's cloaca when she passes by. The spermatophore then slowly enters the female tract, where the enclosed spermatozoa are released to fertilize the eggs.

Sexual Rites in Mammals

In mammals, scents and tactile contact play an essential role in the various forms of sexual behaviour which are encountered. The secretions produced by a bitch in heat attract all the male dogs in the neighbourhood, whereas they show no interest at all if her odour is somehow suppressed. In the past, horse-thieves used to employ a sponge soaked in urine from a mare in heat in order to attract

A stag in rut roars, fights with rival males (above) to drive them off, and continually herds the hinds in order to maintain the cohesion of the group.

In the great crested grebe the nuptial ceremony is particularly ritualized. Strange ballets, with the male and the female performing identical movements, can be seen on lakes and large ponds during the breeding season.

stallions that they wished to steal from a meadow. Chemical substances contained in the secretions of females and exerting an olfactory influence on males during the breeding season are known as pheromones. However, although a female may be in oestrus (heat), as indicated by the pheromones she is producing, and therefore theoretically receptive to the male, a certain state of physical excitation must be aroused in her before she will permit copulation to take place.

In rabbits, the male excites the female by leaping over her while spraying her with a stream of urine. It is the combination of these two stimuli which renders the female receptive to the male for mating. In the large herbivores, a male will perform in front of the female a series of movements which will allow her to recognize him as a member of her own species and also make her receptive. This form of behaviour has been particularly well observed with large African antelope species, in which it is very ritualized. When a female Uganda kob approaches the territory of a male, the latter begins to paw the ground, displaying the black stripes which are present on the forelegs, and then raises his head so that the female can see the white patch on his chin. Subsequently, the male excites the female by touching her flanks repeatedly with one of his forelegs. If the female moves off and leaves the male's territory, he will follow her to his territorial boundary, but will not go beyond it even if a neighbouring male begins to paw the ground as the female approaches. In the Uganda kob, as with virtually all ungulates (hoofed mammals), the ritualized nuptial parades which are performed emphasize secondary sexual characters such as coloured patches, stripes of varying intensity, markings on the rump, tufts of hair, tailtufts, horns or antlers. For other orders of mammals, with the exception of the primates, very little is as yet known about the performance of ceremonies as a prelude to mating. In the cat group (felines), prenuptial displays consist largely of vocalizations, as most people will have noticed with the domestic cat. In insectivores and small rodents, sexual behaviour is confined to a long chase in which the male pursues the female while sniffing at and lifting up her hindquarters. It would seem that the action of hormones alone is sufficient to ensure the normal performance of reproduction.

In theory, this explanation is correct, since without the hormones the behaviour would certainly not occur. But the activity of the pituitary gland which secretes the main

The common frog (below) provides a good example of external fertilization with copulation. The male clasps the female and releases his sperm as she lays her eggs. In the copulatory encounters of dragonflies (right), the male (blue) seizes the female with terminal pincers on his abdomen. She, in turn, curves her abdomen forwards to contact the appropriate spot on the ventral surface of the male's abdomen.

When breeding, male tortoises (above) and turtles perform drawn-out ceremonies which sometimes produce identifiable sounds. Despite the fact that tortoises are famous for their slowness, a male can follow a female with surprising rapidity. When he has caught up with the female, he will strike her carapace with rhythmic blows. Centre: a carp. Lower left: a moray eel. Lower right: a sea-horse. In the sea-horses, the male takes care of the eggs, which he keeps in a ventral pouch until they hatch.

hormones must itself be stimulated by behaviour. There is thus a feedback relationship which is clearly reflected in the overall assemblages of postures and displays collectively referred to as the "nuptial parade". Unfortunately, the nocturnal habits of most mammals make observation of mating displays extremely difficult. It is for this reason that such behaviour has been best studied in birds.

The Nuptial Parades of Birds

Although nuptial parades can be observed in various forms in a wide variety of animals, they are most highly developed in birds. This is, no doubt, due to the fact that birds exhibit very elaborate behaviour in general, combined with bright colours, various forms of adornment and many striking vocalizations. In addition, their ability to move through three dimensions, thanks to their wings, permits them to utilize their external characters more effectively as stimuli.

As a general rule, the nuptial parades of birds are performed by the males, though it has already been noted that there are a number of

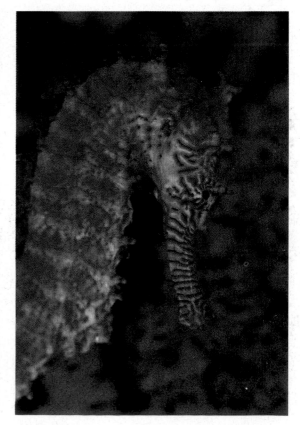

The female curlew (above) seen on its nest in damp meadowland. In this species, the prelude to mating takes place both in the air (with the male flying vertically upwards while producing loud trills) and on the ground (with the male circling around the female and performing invitation displays).

The copulation of large marine mammals also involves complex ceremonies. Dolphins carry out spectacular aquatic ballets. The beluga, or white whale (below), emits a series of high-pitched calls which can be heard 300 metres (900 feet) away.

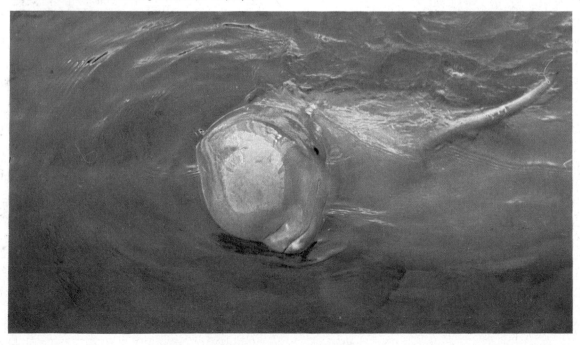

Marine turtles return to coastal areas to lay their eggs in sand where the heat of the sun will assist development of the embryos. When they hatch, the young turtles (below) scurry over the sand to reach the water as fast as possible. They suffer considerable losses during their first excursion, since numerous predators are attracted by the hatchlings. These young snakes emerging from the egg (right) will be as active as their parents as soon as they escape the soft, parchment-like shell in which they have developed. Facing page: spider hatchlings. In many spider species the female encloses her eggs in a kind of pouch which she spins round them to provide support.

species with inverted sexual behaviour where the females take on this role. In addition, in a number of large-bodied bird species where there are no great differences between males and females in their plumage patterns, "mutual" nuptial parades are often observed, with the male and female simultaneously performing the same behaviour patterns. This is the case with grey herons, with great crested grebes and with laughing gulls, to cite only three notable species which are to be found in Europe. However, not all bird species in which the two sexes have an identical external appearance engage in reciprocal nuptial parades. In the starling, the wren and several pigeon species it is the male that performs the nuptial parades alone. According to the species concerned, nuptial parades can take place on the ground, on water or in the air. The contest of the ruff takes place in the middle of an arena on meadowland. The mallard and grebes carry out their preludes to mating on the water. The black kite performs spectacular aerial ballets during the mating season. Usually the primary function of nuptial parades in birds is the display of the male's plumage to the female. A male sparrow adopts special postures to display his black throat, while the redstart spreads his tail and opens his beak wide to display the bright yellow colour of his throat. In many species nuptial parades are accompanied by a variety of different displays, some of which are likely to amaze a casual observer. Good examples are the mutual preening performed by jackdaws, the caresses and "kisses" exchanged by doves and pigeons, the offer of a "gift" of fish among kingfishers and symbolic feeding of the female by the male robin. All these special behaviour patterns serve as an outlet for aggressive tendencies and reduce the female's fear of the male.

Courtship in Other Animals

In frogs and toads, the various kinds of croaking and chirping calls produced are known to be involved in sexual behaviour. The concerts of calls reach their maximum intensity at the time of egg-laying and gradually decrease over the succeeding summer. But there is no real nuptial parade in frogs or toads. On the other hand, although they do not actually copulate, urodeles — such as newts — exhibit special male colour patterns during the breeding season which are accompanied by movements of the tail that seem to serve a display function. In the slow-worms and in lizards generally, males engage in violent fights. In fact, all reptiles exhibit fairly complex sexual behaviour reminiscent of true nuptial parades. In turtles and tortoises, reproduction is also accompanied by fighting between males and noisy pursuits with the animals banging their carapaces together. The males of Hermann's tortoise even injure one another quite seriously during such encounters.

The Sexual Behaviour of Insects

Niko Tinbergen and his students spent many months observing the nuptial dance of a

diurnal butterfly which is very common in Europe, the grayling. During the breeding season, the male waits unmoving among the foliage for females to pass by. As soon as he spots a female, he sets off on a whirling flight in which he circles around the female, coming closer and closer until she is obliged to land. The whole process looks rather like an aerial engagement between two fighter-planes. As soon as the female has landed, the male lands beside her and begins to walk around her, spreading his wings every time he passes in front of her. His wings in fact bear special scent-producing areas which evoke the female's interest. She approaches the male little by little and then passes her antennae (feelers) between the male's wings as if to "sniff" at them. After being "seduced" in this way, she allows the male to mate with her.

In order to find out more about the way in which the first phase of this nuptial dance took place, Tinbergen carried out a number of ingenious experiments by dangling from a cane rod a whole series of artificial "females" made out of paper. He had been struck by the fact that male graylings would take off not only in response to females of their own species, but also in response to females of other species, to dead leaves carried on the wind and to small flying birds. He had even seen male graylings following their own shadows.

From the first series of experiments using artificial females of different shades ranging from black to white, Tinbergen established that, although the males would follow all of them, they preferred the darker ones. Subsequent experiments using various geometric forms — circles, squares and rectangles — showed that they were all equally attractive. Further experiments conducted by Tinbergen and his team showed that a male grayling would respond twice as well to dummy females placed at a distance of ten centimetres (four inches) as to those placed one metre (three feet) away, and that an undulating, dancing

flight of the female was more attractive to the male than straight line flight.

The large stag-beetle of Europe is a familiar sight. It takes its name from the extremely well-developed "antlers" formed by the mandibles in the male. These cumbersome structures are more impressive than they are dangerous, although the males do use them in the fights which take place for the possession of females, and it is only natural to suspect that they might have some other function. The Hercules beetle of tropical South America, which is the largest beetle in the world, belongs to a family closely related to that containing the stag-beetle. In this species, the male is at least twelve centimetres (five inches) long and bears an enormous forward-directed horn which accounts for its alternative name of "rhinoceros beetle". This horn is not really involved in fighting between males as such, but a number of observers have seen it used in a curious fashion to carry off the female after a fight. In fact, it is not

In colonies of social insects, reproduction is carried out by a single female, the queen. Sterile females, known as workers, take care of the eggs and the larvae. Above: bees on a wax comb. Below: a nest of a French wasp belonging to the genus Polistes. Facing page: a colony of caterpillars from butterflies of the genus Malacosoma.

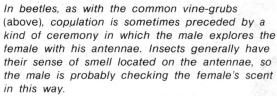

In beetles, as with the common vine-grubs (above), copulation is sometimes preceded by a kind of ceremony in which the male explores the female with his antennae. Insects generally have their sense of smell located on the antennae, so the male is probably checking the female's scent in this way.

Above: a caterpillar of the monarch butterfly on the leaves of its food-plant Asclepia. The striking pattern on its body acts as a warning to insectivorous birds, which avoid eating the caterpillar. Centre: the nymph of this butterfly. Below: the monarch just about to emerge from its chrysalis.

necessarily the winner of the fight who carries off the female, since a male pushed into the vicinity of the female by a violent thrust during a fight will immediately pick her up and carry her away on the horn protruding from his thorax.

Bumble bees (above), which are large hymen-
opterans characterized by their hairy coats and
noisy flight, are social insects. However, their
colonies are much smaller than those of honey-
bees and they contain, in addition to a big found-
ing queen, several smaller females which are also
fertile. The hornet (below right) is a formidable
social wasp which constructs its nest in hollow
trees, beneath rafters and sometimes in the
nesting-boxes provided for birds.

Birth and Hatching

Family Relationships

It is to be expected that the relationships of animals within family groups will be even more complex than those that have already been described for the interactions between a pair of animals of opposite sex. The parents of a family must find a shelter where the female can give birth or lay her eggs, possibly find food for the offspring, and protect them from attack by predators.

Precocial and Altricial Young

The degree to which these three duties of the parents are developed can vary greatly from one animal species to another. Let us begin by considering the parental behaviour of birds, which is the best-known because of the relative ease with which it can be observed.

A young pigeon (squab) is completely naked when it hatches from the egg. It is blind, completely unable to move and therefore unable to feed itself. In contrast, within a few hours of hatching a duckling can accompany its mother in the water and is already able to chase small insects on the water surface. The pigeon squab is said to be altricial: it is entirely dependent upon its parents and unable to live outside the sheltered environment provided by the nest. The duckling is said to be precocial: it abandons the nest as soon as its down feathers are completely dry, just a few hours after hatching.

Since the birds are derived from reptiles and since the latter have precocial offspring, it seems likely that this pattern of development is the most primitive and that the altricial condition represents a more recent stage of evolution. After hatching, altricial birds exhibit the instinctive behaviour which is required for growth and survival. They open their gapes wide to incite their parents to provide them with a continuous supply of food. Because of their initial lack of feathers, they have to remain in the warm nest (which may be specially lined) for the first few weeks of their existence. Adult birds which produce altricial young must therefore construct an appropriate nest and then feed their offspring incessantly from hatching to independence.

Precocial birds, by contrast, emerge from the egg at a more advanced stage of development. Their eyes are open, they have a thick coat of down feathers, and many of them are able to find food unaided almost from the moment of hatching. As a rule, their parents construct only a rudimentary nest, if they construct a nest at all. It should also be noted that altricial offspring require greater protection and defence from predators than do precocial offspring, which are able to escape by fleeing under their own steam. This divergence is reflected in the behaviour of adults with respect to their offspring and, as a result, in the relationships existing within the family group.

Nest Building

The first birds probably followed their reptilian

ancestors in laying their eggs in crevices in the ground or in tree-hollows and covering them with sand, earth or leaves. As the birds evolved, those species which became adapted for life in wetlands and swamps developed the behaviour of seeking out small patches of dry land among the reeds. Subsequently, nest-building above ground-level could have emerged as a response to flooding and rising water levels.

Not forgetting that there are some birds, such as the cuckoo, which do not build a nest and do not incubate their eggs, we can recognize five kinds of nest among the birds:

— ground nests, such as those of pheasants, partridges and curlews.

— floating nests, such as those of the mallard, the grebes, moorhens and coots.

— nests in crevices, either in a ready-made hollow (sparrow-owl, coal-tit) or in a hollow made by the parent bird itself (kingfisher, sand-martins, bee-eaters and woodpeckers).

— suspended nests (wrens, sedge-warblers).

— walled nests (house-martin and common swallow).

Some birds do not build a nest at all, either because they parasitize the nests of other birds (e.g. the cuckoo), or because their offspring are precocial (e.g. the emperor penguin), or because they install themselves in nests constructed by other species, as is the case with the redstart, which readily uses the old nests of blackbirds or thrushes.

The collection of nest-building material is sometimes a cause of competition between

The flying squirrel (below) nests in tree hollows where its young are more easily protected against predators.

Most bird species which nest in trees construct a cup-like nest (below). However, there are a number of birds, such as the magpie, which do build a roof over the nest.

species. Tits, for example, will often rob the lining from crows' nests.

The Lairs of Mammals

As with birds, it is possible to distinguish two categories of mammal offspring. Those in the first category are entirely dependent on their mothers for the early part of their lives. Examples are young rabbits and kittens which are born blind and naked. Those in the second category are sufficiently well-developed at birth to move around independently soon afterwards.

This is the case with many ungulates (hoofed mammals) such as the roe-deer and the chamois.

The nature of the lair where the female gives birth is largely dependent on this difference between the two kinds of offspring. Roe-deer and hares simply give birth on the ground, whereas the female rabbit digs a special burrow (the stop) which she carefully closes off after each visit to the pups to suckle them.

The shelters which mammals build to house

In order to build its den, the beaver (above) will begin by accumulating a pile of branches close to the water. It then digs a burrow into the bank below the water level. This eventually opens out inside the wood pile, which is then gnawed into appropriate shape. In addition to the family den which houses a pair and their offspring, there are other dens which are temporarily occupied, notably that which the male uses when he is separated from the female while she is giving birth.

and protect their young show just as much variety as do birds' nests, though it should be noted that these shelters generally serve as refuges for adults as well. Some mammals, for example most bats, make use of natural hollows such as rock fissures, tree-hollows and caves. Others dig burrows of varying depth and complexity (mole, rabbit, vole), or take over and enlarge burrows which have already been dug by another mammal (fox). In certain cases, nests are constructed above ground-level and they may look rather like birds' nests, as is the case with those of the harvest mouse and of some squirrel species. Finally, there are a few mammals which build veritable edifices, the most striking example being the beaver's den.

Parental Care

In birds, the elaborate parental care devoted to the offspring is not limited to the provision of food. Altricial nestlings are initially unable to maintain constant body temperature unaided and they can only manage this after a period of growth. During this transition period, the parents must warm up the offspring if they are cold and protect them from excessive heat. Most birds keep their offspring warm by squatting on top of them and they can spread their wings if necessary to protect them from intense sunlight. Some bird species will defend their offspring with enormous courage if they are attacked by a predator, while others will try to distract the predator. For example, in order to lure cats, dogs or foxes away from its nests on the ground, the ringed plover limps along and clumsily flaps one wing as if it were broken.

The male may sometimes play an active role in the rearing of the young in mammals. In both wolves and foxes, the male provides a great deal of the food for the family while the young are being reared. In other cases, the male will defend his offspring against intruders, as can be observed with the roe-deer, for example. Even with species in which the male has no further relationship with the female after mating, the males sometimes show a certain degree of interest in their offspring.

Following a gestation period of three months, the tigress (above) will give birth to one to five cubs which she will suckle for about six weeks.

After a gestation period of between 200 and 210 days, the young of the female grizzly bear (left) are born in winter at the bottom of a den in which they pass the most difficult period of the year huddled together with their mother. At birth, the offspring weigh only 250—300g (9—11oz).

Learning and Independence

Numerous observations and experiments, particularly those conducted with mammals and birds, have shown that the first experiences in an animal's life are often determined by the behaviour of the adults, reflecting their characteristic conduct.

As far as embryologists are concerned, the development of behaviour begins even before birth. The embryos of many bird species, for example, learn to recognize their mother and to call to her while they are still enclosed in the egg. It is certain that the young bird in such cases begins to call before it has hatched and that its calls establish the first contact with its parents. As a rule, the parents do not help the young to escape from the egg, though this has been observed with the water-rail and the stone-curlew.

In mammals, the embryo is much more isolated from the external environment, but it depends more directly on the physiological condition of the mother. If the mother is subject to violent emotional disturbances, the behaviour of her offspring may subsequently reflect this. With laboratory rats it has been observed that offspring born to "stressed" mothers were more timid when they were placed in an unfamiliar environment.

Learning

Soon after a gull chick emerges from its egg, it follows a number of characteristic behaviour patterns. It first stays close to its mother without moving for a while, then — often before its feathers have properly dried — it will begin to peck weakly at the female's beak every time that she approaches the nest. These pecks gradually become more frequent and more precise and the mother responds by regurgitating food. Later on, the chick begins to find its own food. After initially pecking at a wide range of different objects, it will eventually avoid those which are inedible and only eat those which represent suitable food. This is typical of the way in which behaviour develops in many animal species. Some actions are performed adequately or even perfectly on the first occasion, while others develop gradually. It is tempting to conclude that behaviour of the first kind is innate while that of the second kind is learned. In fact, the situation is far more complicated than this, since it is necessary to distinguish between internal and external causes. This can be illustrated with the following example.

A chick which has just hatched from an egg incubated by its mother immediately begins to follow her around. But if the egg hatches

When the elephant (right) *sprays itself with water from its trunk, this may serve the serious function of removing irritating parasites, but it may also be a form of play.*

in an artificial incubator, the chick will follow the first moving object that it encounters after emergence from the egg. From this it can be concluded that the chick is "programmed" to follow, but that the "programming" is incomplete, since if the unknown mother is absent anything may be followed in her place. For a chick to follow its mother, the innate response exhibited immediately after hatching must be completed by contact with the external environment. In other words, the individual must adjust its behaviour according to its own experience. Many other behaviour patterns of young animals develop in a similar fashion as a result of contact with the external environment and the manner in which it responds to the contact. Although birds do not have to learn the basic essentials of flight, fine tuning of this ability doubtless depends upon experience. When birds are observed emerging from the nest, they are often found to have difficulty in descending through the air and they tend to fly higher and higher, in some cases being carried away by the wind. It is also through learning that young seagulls eventually acquire the ability to land against the wind.

Play

For many animal species, particularly for mammals, play is at least as important as the mother's behaviour for determining the future social behaviour of the developing young. Play behaviour in the young can, in fact, depend quite heavily on maternal behaviour.

Bear cubs (above and left) *remain with their mother for more than a year after birth. During this time, they learn to seek out their food and to protect themselves. When their first winter comes, they go to the den with the mother bear and share it with her until the arrival of Spring.*

Harry F. Harlow showed that a young rhesus monkey deprived of its mother from birth onwards was unable to engage in play with other young monkeys provided as companions. The deprived youngster always proved to be indifferent, or even aggressive, towards its companions, and on attaining sexual maturity it was unable to perform normal copulation. By enabling a young animal to expand the contacts it has with its environment in general and with its siblings in particular, play increases the range of experience and rounds out the process of learning. In physical terms play also favours better motor co-ordination, improves respiratory functions and develops the musculature, thus preparing the growing animal more effectively for adult life. For this reason, play behaviour is particularly developed in predatory mammals. The cubs of lions, wolves and foxes, which are all rather clumsy and ungainly in their initial scuffles, rapidly acquire greater agility in both attack and avoidance. Chasing play also enables young gazelles to develop a capacity for high-speed running which improves their chances of escaping the jaws of their many predators.

Many of these observations concerning the importance of maternal behaviour and of play also apply to human infants, to the extent that we depend in similar ways on the effects of experience.

Just like a domestic cat, the female leopard (above) carries her cubs by the scruff of the neck when she feels they are threatened. A large number of mammal species and even a number of bird species exhibit this kind of infant transport.

Independence

Weaning and the attainment of independence which follow represent a difficult period for young mammals, since many of them are still ill-equipped for survival unaided. For this reason, they will often continue to live in a family group for a varying period, depending on the age at which sexual maturity is reached. Although growth does not necessarily cease at sexual maturity, attainment of this stage represents the end of the juvenile period in behavioural terms. As soon as they have reached this stage of their development, the young animals become competitors with respect to their parents. This is particularly true of solitary species and of those which live in permanent pairs, as is the case with bears and beavers respectively. The sexually mature offspring of such species are mercilessly chased away, receiving the same aggressive treatment from their parent as would be meted out to strangers of the same species.

Ringtail lemurs (above) are diurnal prosimians of Madagascar which live in social groups of up to two dozen individuals. Among big carnivores such as the lion (left), adults often join in play with the young, as this lioness is doing with her cub.

Social Organization and Animal Groups

Contrary to popular opinion, it is not true that animals can be split into two distinct categories of solitary versus social. There are no truly solitary animals, remaining alone for their entire lives. A minimum degree of relationship must be established when the sexes meet for reproduction, however short-lived this may be. An extreme case is provided by the Eastern American chipmunk (ground squirrel), in which the female and the male resume their mutual antagonism just a few minutes after mating. In addition, in mammals a relationship is established between a mother and her offspring during the suckling period. This relationship is undoubtedly social in nature, even if the family ceases to exist as soon as the young leave the nest, which is true once again of the Eastern American chipmunk.

Group Living

Nevertheless, we shall be considering here only those relationships between individuals living in permanent groups characterized by a proper hierarchical pattern of social organization and justifying the term animal societies. Examples of such groups are the herds of African ungulates, the schools formed by many fish species, the migratory flocks of birds and the colonies formed by ants and termites.

The advantages of living in social groups are obvious. Leaving aside those species in which the sexes live separately and which could not survive if males and females did not at least encounter one another during the breeding season, thus requiring an elementary degree of cooperation between them, there are numerous examples of the survival value of group living. Caterpillars living in colonies provide warmth for one another and thus accelerate the developmental process. Seagulls which gather together in colonies during the breeding season can form a united front to protect their nest against predators. (Any members of the colony which lay their eggs before or after the others lose many eggs.) Among lions group-hunting increases the success rate, and so on.

As is shown by these few examples, social animals do not simply gather together in groups — they undertake joint activities which are governed by their overall contribution to the well-being of group members. In addition, there is often a real division of labour between the members of a particular group. The organization of a beehive represents an extreme case, with the queen, millions of worker-bees and, at certain times of year, males, all playing their specific role in the colony. The activities of the worker-bees are themselves subdivided according to age.

Animal societies are therefore often characterized by a hierarchy which is established by a process which, from a human point of view, would be branded as "antisocial", since it depends upon the principle "might is right". Individual group members establish relation-

One of the most obvious advantages of group life is better protection against attack from predators. Above: a colony of fur-seals on the Pribyloff Islands. Left: a herd of African buffalo.

When a troop of baboons is on the move, the young ones and the dominant males stay in the centre of the troop with the adult females, while lower-ranking adult males and juveniles of both sexes keep to the periphery. The peripheral animals are the ones that raise the alarm if there is any threat to the group.

A herd of elephants is composed of a number of family units of a special kind, each containing an adult female and her young of different ages. The overall herd is led by an adult female, usually one of the oldest in the group. Adult males form a separate herd which keeps its distance from the female herd. There are also a number of adults which lead a solitary existence.

ships of "dominance" and "submissiveness". The naturalist Schjelderup-Ebbe noted in 1913 that a social hierarchy exists among the chickens in a chicken-coop. The individual chickens in the coop adopt aggressive or submissive roles according to the number of pecks given and received. Subsequently, several research-workers studying domesticated animals or captive animals discovered similar relationships within groups.

Similar hierarchies undoubtedly exist in a great number of animals species living under natural conditions. Social hierarchies in general are always somewhat unstable and they change over a period of time according to the age and physiological condition of individual group members.

One of the first examples which comes to mind when talking about social life in mammals is that of the beaver. According to the well-established, but erroneous, traditional view, these rodents are supposed to live in very large colonies. Recent studies have in fact shown that the beaver colony, that is to say a group of individuals living in an identifiable territory (whose boundary is marked by scent-gland secretions), using the same food resources, and making use of the same set of dams, is nothing more than a family group. At its maximum, such a group will consist only of a pair, their young from the present year and older offspring from the previous year. Given the fact that the maximum number of offspring produced is three per year, the "colony" contains no more than eight individuals, far less than imagined by early authors.

It now seems likely that marmot colonies are also based on a family group, differing sharply from the gregarious groups seen with most marine mammals, especially seals and sea-lions, and with ungulates and monkeys, where genuinely solitary species are exceptional. Elephants, wild horses, hippopotamuses, wild boar, most bovids, giraffes and most cervids — to cite only the best known examples — live in social groups in which there is an established hierarchy based, as described above, on dominant-subordinate relationships,

however subtle these may be. The same applies to many carnivore species, such as the wolf and the hunting dog. But such social hierarchies would be difficult to understand in the absence of some communication system between the members of the group, providing a "language" involving recognizable signals. Although a lot has been written, including a great deal of nonsense, about the languages used by animals to "speak" with one another, it is only over the past fifty years that we have begun to understand these signal systems and the part they play within animal social groups.

Animal "Languages"

Animals communicate with one another by means of a series of sign stimuli which evoke specific responses from other members of the same species. These sign stimuli can be classified into three groups: acoustic, olfactory and visual.

Acoustic signals may have a vocal origin (e.g. the calls of mammals and birds) or they may be produced by other organs, as is the case

with the stridulation of grasshoppers and cockroaches or with the musical rattling noise produced by a sea-horse when it shakes its head. According to the species and the exact nature of the sounds produced, these signals can have a variety of functions. They can serve as warnings during territorial defence and aggression; they can also express fear, give the alarm or act as a summons during reproduction. Howler monkeys, for instance, have a repertoire of fifteen to twenty different vocalizations which elicit specific behavioural responses from other members of the group: gurgling noises, loud howls, grunts, groans and so on.

The calls produced by animals are not necessarily audible to the human ear. This is particuarly the case with high-frequency sounds such as those emitted by dolphins. The various sounds that each species produces seem in most cases to be determined by heredity, as in mammals generally and in most other animals.

It has already been noticed, however, that

Wolves (above) *form packs which, like the packs of dogs which can sometimes be seen roving around the outskirts of big towns in Africa and Asia, exhibit very clear social hierarchies. The pack is led by the strongest male, and this dominant wolf also determines the hunting area and the strategy adopted for capturing prey. The other members of the pack display their subordinate status by a whole series of signals based essentially on the positions adopted by the ears and the tail. The dominant male, for example, carries his tail in the air and pricks up his ears. The repertoire of sign stimuli is completed by the raising of the hair and a number of facial expressions.*

In baboon troops (facing page), the females and their young (right) are continuously provided with protection. The adult males, when united together, are able to face up to a big cat and drive it off. The baby chimpanzee (above) also lives in a social group.

some birds have to learn their songs.

Among the non-vocal auditory signals, special mention must be made of those which arise during displays performed by an animal, such as the drumming that certain woodpeckers produce on tree-trunks, the tail-beating performed by beavers on the water surface, and the drumming produced by rabbits with rapid stamping movements of their hindlimbs.

Olfactory signals are also very important in animal communication, particularly among mammals and among insects (as we have seen with the grayling moth). However, human-beings are limited in their ability to appreciate this form of communication because of their own poor sense of smell. Scent is not always of advantage to an animal, however. The scent traces left behind by herbivorous mammals and birds can attract the attention of predators, as can be seen with a hunting dog when it has picked up a trail. All the special odours of animals are produced by glands on the body surface which give each species its characteristic smell, as with pigs or cows. In addition, each individual within a species has its own special odour. A beaver, for example, is able to tell from a deposit of castoreum whether it has been produced by a member of the family group or by a stranger.

Scent glands can be located in a wide variety of places on a mammal's body. In the ungulates alone, there are glands near the eyes, in the groin, at the base of the horns, at the back of the head, on the tail, on the prepuce, on the hocks, on the soles of the feet or at the bases of the hooves. The secretions from the glands may be deposited in the course of normal bodily contact (if on the feet or on the hocks, for instance) or by a special act of rubbing against branches and other objects. In some species, such as the beaver, the scent is sprayed by the animal as it squats on a small hillock of mud or sand.

Visual Signals

Mammalian visual signals have been far less intensively studied than those of birds and fish. In many cases, visual signals involve special colour patches which the animal displays in a particular situation. The most striking example is provided by the mule deer and the American pronghorn antelope, both of which have a patch of long white erectile hairs around the anal region. The white hairs

Adélie penguins (below right) are essentially social and gregarious creatures. During winter, they gather in enormous flocks on the icy shores of Antarctica. Some observers have estimated that up to half a million of these penguins may gather on the 200 hectare (500 acre) area of Adélie Land. Most cormorant species (below left), of which there are about thirty altogether, also nest in colonies. It is best to steer clear of such colonies because of the extraordinarily foul odour they produce and the hordes of flies they attract.

are raised, as is the tail, whenever danger threatens. This alarm signal, which can be seen from a long distance away, immediately provokes the entire herd to take flight. Closer to home, it is well-known that the male wild rabbit has a similar patch of white hairs with which he attracts the attention of the female during the mating season.

With a certain number of mammal species, particularly in the dog family (Canidae) and in the cat family (Felidae), specific signals are given by the positions adopted by the head, ears and tail. In wolves, a high-ranking animal keeps its ears pricked up and its tail raised. Flattened ears betray fear or suspicion. Wrinkling of the nose and retraction of the lips to expose the teeth represent a threat display. The movements of the tail are more difficult to interpret, however. Side-to-side wagging of the tail while it is held aloft acts as an intimidating display, whereas the same wagging movement when the tail is low indicates complete submission. This is the posture that is frequently adopted by a dog when greeting its master. Finally, if the animal is very uneasy indeed, it will tuck its tail between its legs and hold it beneath the belly.

We must also consider the fact that some signals are detected with specially adapted organs found only in certain animal species. For example, the sonar used by bats for echo-location, the lateral line system of fish (which is sensitive to underwater vibrations), the antennae of various insects and the whiskers adorning the snout of many different mammal species.

The Social Monkeys and Apes

The hierarchical social groups found in monkeys and apes have been intensively

In colonial bird species, social behaviour seems to be involved in the tight synchronization of every aspect of reproduction. This is certainly true of the pink flamingo (above).

studied not only by naturalists but also by psychologists and anthropologists who have ·attempted to gain insights into our own social behaviour in the process. Among the monkeys and apes, there is a considerable diversity of patterns of social organization.

Some species, such as the gibbons, live in family groups occupying a well-defined territory which is defended against neighbouring groups. Each group is composed of the two parents, the most recent infant and a variable number of older offspring. As these lesser apes produce only a single infant once every two or three years, the family remains extremely tight-knit.

Hamadryas baboons, rhesus monkeys and Indian langurs are all polygamous and live all the year round in harem-type groups with a variable number of adult females, controlled by a dominant male who must continually defend his rank against competition from younger males attempting to climb the social ladder.

The howler monkeys of the tropical forests of Central and South America form troops which usually contain about three adult males, eight adult females, three infants which are still being suckled by their mothers and four immature juveniles. There is no clear-cut hierarchy among the males of a troop; they co-operate together to lead the social group and to defend it against the incursions of neighbouring troops.

The anthropoid apes (or great apes) have only recently been studied under natural conditions in order to provide an accurate picture of their social behaviour. In addition to the studies conducted by George Schaller and Dian Fossey on lowland and mountain gorillas, a long-term field study of chimpanzee behaviour has been conducted by Jane Goodall and her co-workers in East Africa. In chimpanzees, experienced old males lead the group and keep a careful look-out for danger.

The Dance of the Bees and Social Insects

Foremost among the social insects, the bees with their complex behaviour and social

There are a number of different weaver-bird species which construct nests suspended from trees (above) using dried grass. The South-West African social weavers, which are close relatives of the European sparrows, build enormous communal nests containing a large number of individual pairs, each with their own nest entrance.

organization within the hive have always fascinated human observers. This interest in bee behaviour was undoubtedly sharpened by the fact that these insects provide a valuable food-source and that detailed knowledge of their habits could therefore be of great practical value. However, it is only relatively recently that really detailed observations, conducted by Karl von Frisch and the French scientist Rémy Chauvin, have permitted a clear understanding of the mechanism governing the social behaviour of bees. It is impossible here to cover the innumerable experiments conducted by these two research-workers, or effectively review their discoveries. But the most extraordinary observations are surely those made by von Frisch in discovering the manner in which bees communicate with one another with their "round-dances" and their "tail-wagging dances".

Whenever a bee returns to the hive with a load of nectar, this is first deposited and then the bee performs a round-dance which arouses other bees. The more dancing bees there are, the greater the number of other bees which crowd around them. Their dancing informs the other members of the hive that they have found an abundant source of food. It might be expected that at the end of the round-dance the other worker-bees would simply follow those that discovered the food source. However, the distance of the source is actually indicated by a different kind of dance, the "tail-waggling dance", which is performed in a figure of eight. The frequency of the tail-waggling motion corresponds to a sliding scale of distances (or, more exactly, to the time taken to fly over them) and the watching worker-bees are able to register and interpret this information with remarkable precision. Von Frisch was even able to establish that if there was a head-wind the dancing bee would

Many fish species live in tightly packed shoals which may contain an enormous number of individuals, all moving together as if they were under a single command. A shoal of trout (above) provides a typical example.

In South American forests leafcutter ants (right) *cut small pieces from leaves and then carry these to their nest. Inside the nest, in special chambers, other workers mix these leaf fragments with waste products to produce a kind of compost which enhances the growth of fungi. The spores of the fungi provide a special food-source for the ants.*

indicate a greater distance (or flying time) than would be the case in calm weather. This "tail-waggling dance" also indicates to the nectar-gathering bees the location of the food source by means of its orientation in relation to the sun.

The behaviour of ants and termites has not yet been studied in all its detail. The most striking aspect of the organization of these insect societies is the manner in which the problem of division of labour has been resolved in the course of their evolution. It now seems to be emerging that the classic division into a fertile queen, workers and soldiers proposed by early observers is not as clear-cut as was originally thought.

All the four thousand known ant species are social forms, whereas there are some bee and

Many ant species collect honey-dew from aphids and increase the secretion of this substance by caressing the aphids with their antennae (right). *The Amazonian army ants* (far right) *form a long column which moves along devouring everything in its path.*

The organization of a termite colony is very similar to that of an ant's nest. Each colony is founded by a pair of winged termites. The queen lives together with the male in a central cell where their only task is the production of eggs. The workers provide the developing larvae with food while the bigger and stronger soldiers defend the colony against attack. Above: a group of termite mounds in the African savannah. Left: a worker-termite. Below left: winged termites, just about to embark on their nuptial flight, alongside a number of workers and immature termites.

wasp species which are solitary in habits. Termites, also known as "white ants" although they are actually zoological relatives of the cockroaches, exhibit a pattern of social organization quite similar to that of the ants. That is to say, their colonies exhibit a division into castes.

Both ants and termites exhibit the exchange of food (trophallaxy) among colony members, and this plays an important part in the organization of their social life.

Animal Society in Literature

The Seagull

from **Jonathan Livingston Seagull** by Richard Bach

By sunup, Jonathan Gull was practising again. From five thousand feet the fishing boats were specks in the flat blue water, Breakfast Flock was a faint cloud of dust motes, circling.

He was alive, trembling ever so slightly with delight, proud that his fear was under control. Then without ceremony he hugged in his forewings, extended his short, angled wingtips, and plunged directly toward the sea. By the time he passed four thousand feet he had reached terminal velocity, the wind was a solid beating wall of sound against which he could move no faster. He was flying now straight down, at two hundred fourteen miles per hour. He swallowed, knowing that if his wings unfolded at that speed he'd be blown into a million tiny shreds of seagull. But the speed was power, and the speed was joy, and the speed was pure beauty.

He began his pullout at a thousand feet, wingtips thudding and blurring in that gigantic wind, the boat and the crowd of gulls tilting and growing meteor-fast, directly in his path.

He couldn't stop; he didn't know yet even how to turn at that speed.

Collision would be instant death.

And so he shut his eyes.

It happened that morning, then, just after sunrise, that Jonathan Livingston Seagull fired directly through the centre of Breakfast Flock, ticking off two hundred twelve miles per hour, eyes closed, in a great roaring shriek of wind and feathers. The Gull of Fortune smiled upon him this once, and no one was killed.

By the time he had pulled his beak straight up into the sky he was still scorching along at a hundred and sixty miles per hour. When he had slowed to twenty and stretched his wings again at last, the boat was a crumb on the sea, four thousand feet below.

His thought was triumph. Terminal velocity! A seagull at *two hundred fourteen miles per hour!* It was a breakthrough, the greatest single moment in the history of the Flock, and in that moment a new age opened for Jonathan Gull. Flying out to his lonely practice area, folding his wings for a dive from eight thousand feet, he set himself at once to discover how to turn.

A single wingtip feather, he found, moved a fraction of an inch, gives a smooth sweeping curve at tremendous speed. Before he learned this, however, he found that moving more than one feather at that speed will spin you like a rifle ball . . . and Jonathan had flown the first aerobatics of any seagull on earth. He spared no time that day for talk with other gulls, but flew on past sunset. He discovered the loop, the slow roll, the point roll, the inverted spin, the gull bunt, the pinwheel.

When Jonathan Seagull joined the Flock on the beach, it was full night. He was dizzy and terribly tired. Yet in delight he flew a loop to landing, with a snap roll just before touchdown. When they hear of it, he thought, of the Breakthrough, they'll be wild with joy. How much more there is now to living! Instead of our drab slogging forth and back to the fishing boats, there's a reason to life! We can lift ourselves out of ignorance, we can find ourselves as creatures of excellence and intelligence and skill. We can be free! *We can learn to fly!*

The years ahead hummed and glowed with promise.

The gulls were flocked into the Council Gathering when he landed, and apparently had been so flocked for some time. They were, in fact, waiting.

"Jonathan Livingston Seagull! Stand to Centre!" The Elder's words sounded in a voice of highest ceremony. Stand to Centre meant only great shame or great dishonour. Stand to Centre for Honour was the way the gulls' foremost leaders were marked. Of course, he thought, the Breakfast Flock this morning; they saw the Breakthrough! But I want no honours. I have no wish to be leader. I want only to share what I've found, to show those horizons out ahead for us all. He stepped forward.

"Jonathan Livingston Seagull," said the Elder, "Stand to Centre for shame in the sight of your fellow gulls!"

It felt like being hit with a board. His knees went weak, his feathers sagged, there was a roaring in his ears. Centred for shame? Impossible! The Breakthrough! They can't understand! They're wrong, they're wrong!

". . . for his reckless irresponsibility," the solemn voice intoned, "violating the dignity and tradition of the Gull Family . . ."

To be centred for shame meant that he would be cast out of gull society, banished to a solitary life on the Far Cliffs.

". . . one day, Jonathan Livingston Seagull, you shall learn that irresponsibility does not pay. Life is the unknown and the unknowable, except that we are put into this world to eat, to stay alive as long as we possibly can."

A seagull never speaks back to the Council Flock, but it was Jonathan's voice raised. "Irresponsibility? My brothers!" he cried. "Who is more responsible than a gull who finds and follows a meaning, a higher purpose for life? For a thousand years we have scrabbled after fish heads, but now we have a reason to live — to learn, to discover, to be free! Give me one chance, let me show you what I've found . . ."

The Flock might as well have been stone.

"The Brotherhood is broken," the gulls intoned together, and with one accord they solemnly closed their ears and turned their backs upon him.

Jonathan Seagull spent the rest of his days alone, but he flew way out beyond the Far Cliffs. His one sorrow was not solitude, it was that other gulls refused to believe the glory of flight that awaited them; they refused to open their eyes and see.

He learned more each day. He learned that a streamlined high-speed dive could bring him to find the rare and tasty fish that schooled ten feet below the surface of the ocean: he no longer needed fishing boats and stale bread for survival. He learned to sleep in the air, setting a course at night across the offshore wind, covering a hundred miles from sunset to sunrise. With the same inner control, he flew through heavy sea-fogs and climbed above them into dazzling clear skies . . . in the very times when every other gull stood on the ground, knowing nothing but mist and rain. He learned to ride the high winds far inland, to dine there on delicate insects.

What he had once hoped for the Flock, he now gained for himself alone; he learned to fly, and was not sorry for the price that he had paid. Jonathan Seagull discovered that boredom and fear and anger are the reasons that a gull's life is so short, and with these gone from his thought, he lived a long fine life indeed.

They came in the evening, then, and found Jonathan gliding peaceful and alone through his beloved sky. The two gulls that appeared at his wings were pure as starlight, and the glow from them was gentle and friendly in the high night air. But most lovely of all was the skill with which they flew, their wingtips moving a precise and constant inch from his own.

Without a word, Jonathan put them to his test, a test that no gull had ever passed. He twisted his wings, slowed to a single mile per hour above stall. The two radiant birds slowed with him, smoothly, locked in position. They knew about slow flying.

He folded his wings, rolled, and dropped in a dive to a hundred ninety miles per hour. They dropped with him, streaking down in flawless formation.

At last he turned that speed straight up into a long vertical slow-roll. They rolled with him, smiling.

He recovered to level flight and was quiet for a time before he spoke. "Very well," he said, "who are you?"

"We're from your Flock, Jonathan. We are your brothers." The words were strong and calm. "We've come to take you higher, to take you home."

"Home I have none. Flock I have none. I am an Outcast. And we fly now at the peak of the Great Mountain Wind. Beyond a few hundred feet, I can lift this old body no higher."

"But you can, Jonathan. For you have learned. One school is finished, and the time has come for another to begin."

As it had shined across him all his life, so understanding lighted that moment for Jonathan Seagull. They were right. He *could* fly higher, and it *was* time to go home.

He gave one last long look across the sky, across that magnificent silver land where he had learned so much.

"I'm ready," he said at last.

And Jonathan Livingston Seagull rose with the two starbright gulls to disappear into a perfect dark sky.

The Musicians of Bremen

by the Brothers Grimm

ONCE UPON A TIME there was a donkey who had worked faithfully for many years. However, his strength began to fail and his master wanted to get rid of him, so the donkey decided to run away to Bremen. He felt sure that he could get work there as a perfoming musician. When he had gone a little way, the donkey found a dog lying panting on the road as though worn out.

"Why are you panting so, Growler?" asked the donkey.

"Because I am old," answered the dog, "and can no longer keep up with other dogs, my master wanted to kill me. I ran away, but I have no idea how I can earn a living."

"Come with me to Bremen," said the donkey. "I am going to become a performing musician. I shall play the lute and you can beat the kettledrum".

The dog agreed, and on they went.

Soon afterwards, they found a cat sitting gloomily at the side of the road.

"What is bothering you, Whiskers?" asked the donkey.

"Who can be cheerful when his life is in danger?" said the cat. "I am getting old and would sooner sit by the stove than hunt mice. Because of this my mistress wanted to drown me. I've run away, but I have no idea how I can earn a living."

"Come with us to Bremen," said the donkey. "You have a bewitching voice. You can become a performing musician like us."

The cat agreed and joined them.

Next, the three passed a yard where a cock was crowing desperately.

"What is the matter?" asked the donkey.

"My mistress has ordered the cook to make me into soup," the cock replied.

"Come along, Redcomb," said the donkey. "We are on our way to Bremen. You have a fine strong voice. You can become a performing musician like us."

The cock agreed, and the four of them went on together.

That evening they saw a light burning in the distance.

"There must be a house not far off," called out the cock. When they approached the house, the donkey went up to the window and looked in.

"I can see a table laden with food and drink," he said, "and robbers sitting round it enjoying themselves."

"That place would just suit us," said the cock, and at once the animals began to think how they could drive out the robbers.

"The rest of you could climb on top of me," suggested the donkey. "Then I could put my forefeet against the wall by the window and we could break in and terrify them."

The others agreed, so, when the donkey gave the signal by braying, the cat mewed, the dog barked, the cock crowed and they all smashed the glass in the window.

At this terrible noise the robbers fled. Then the animals climbed in and sat down to dine, eating as though they had not seen food for weeks. Afterwards, they put out the light and each chose a sleeping place. The donkey lay down outside, the dog curled up behind the door, the cat stretched out on the hearth near the warm ashes and the cock flew up to perch on the rafters. Tired out, they soon fell asleep.

The robbers had kept watch from a distance. When all seemed quiet, their chief ordered one of them to examine the house. The robber entered the kitchen to kindle a light. Thinking that the cat's glowing eyes were the coals smouldering in the fireplace, he held a match up to them to light the fire. The cat immediately flew at his face, spitting and scratching. The man was greatly frightened, and tried to escape by the back door. But the dog, who was lying there, jumped up and bit his leg. Then, as the man ran out into the yard, the donkey gave him a powerful farewell kick. Meanwhile the cock, awakened by the noise, crowed at the top of his voice, "Cock-a-doodle-doo!"

Terrified, the robber ran back as fast as he could to his chief, and said, "There is a witch in the house. She spat at me and scratched me with her sharp fingernails. Behind the door stood a man with a knife. He tried to stab me. In the yard lay a shadowy monster. He struck at me with a club. While from high up — like a judgement upon me — a voice cried, "Guil-ty-are-you!""

From that time forward, the robbers dared not go near their den, and the four performing musicians were so delighted with it that they decided to live there for the rest of their days.

Between Two Worlds

from **Born Free** by Joy Adamson

One morning we followed circling vultures and soon found a lion on a zebra kill. He was tearing at the meat and paid no attention to us. Elsa stepped cautiously from the car, miaowing at him, and then, though she did not get any encouragement, advanced carefully towards him. At last the lion looked up and straight at Elsa. He seemed to say, "Don't you know lion etiquette? How dare you, woman, interfere with the lord while he is having his meal? You are allowed to kill for me, but afterwards you have to wait till I have had my lion's share, then you may finish up the remains." Evidently poor Elsa did not like this expression and returned as fast as she could to the safety of the car. The lord continued feeding and we watched him for a long time, hoping that Elsa might regain her courage; but nothing would induce her to leave her safe position.

Next morning we had better luck. We saw a topi standing, like a sentry, on an ant-hill, looking intently in one direction. We followed his glance and discovered a young lion resting in the high grass, sunning himself. He was a magnificent young male with a beautiful blond mane, and Elsa seemed attracted by him. Just the right husband for her, we thought. We drove to within thirty yards of him. The lion looked mildly surprised when he saw his prospective bride sitting on the top of a car, but responded in a friendly manner. Elsa, apparently overcome by coyness, made low moans but would not come off the roof. So we drove a little distance away and persuaded her to get down, then, suddenly, we left her and drove round to the other side of the lion: this meant that she would have to pass him in order to reach us. After much painful hesitation, she plucked up enough courage to walk towards the lion. When she was about ten paces away from him, she lay down with her ears back and her tail swishing. The lion got up and went towards her, with, I am sure, the friendliest intentions, but at the last moment Elsa panicked and rushed back to the car. We drove away with her and, strangely enough, right into a pride of two lions and one lioness on a kill.

This was luck indeed. They must have killed very recently for they were so intent upon their meal that however much Elsa talked to them they paid not the slightest attention to her. Finally they left the kill, their bulging stomachs swinging from side to side. Elsa lost no time in inspecting the remains of the carcass, her first contact with a real kill. Nothing could have served our purpose better than this meal, provided by lions and full of their fresh scent. After Elsa had had her fair share, we dragged the kill back to the handsome young lion who had seemed so friendly. We hoped that if Elsa provided him with a meal he would have a favourable opinion of her. Then we left her and the kill near to him and drove away. After a few hours we set out to see what had happened, but met Elsa already half-way back to the camp. However, since this lion had shown an interest in her, we took her back to him during the afternoon. We found him still in the same place. Elsa talked to him from her couch as though they were old friends, but had plainly no intention of leaving the car.

To induce her to quit her seat, we drove behind a bush and I got out but was nearly knocked over by a hyena who dashed out of his cool retreat, in which we then found a newly killed baby zebra, no doubt provided by the blond lion. It was Elsa's feeding time, so regardless of the consequences, she jumped out of the car on to the carcass. We took this opportunity to drive away as fast as we could and left her alone for her night's adventure. Early next morning, anxious to know the outcome of the experiment, we set off to visit her, hoping to find a happy pair. What we found, was poor Elsa, waiting at the spot at which we had left her but minus the lion and minus the kill. She was overjoyed to see us, desperate to stay with us, and sucked my thumbs frantically to make sure that everything was all right between us. I was very unhappy that I had hurt her feelings without being able to explain to her that all we had done was intended to be for her good. When she had calmed down and even felt safe enough in our company to fall asleep, we decided, rather sadly, that we must break faith with her again and we sneaked away.

Till now we had always given her her meat already cut up, so that she should not associate her food with living animals. Now we needed to reverse our system, so during her midday sleep we drove sixty miles to shoot a small buck for her. We had to go this distance because no one was allowed to shoot game near the camp. We brought her a complete buck wondering if she would know how to open it, since she had had no mother to teach her the proper way of doing it. We soon saw that by instinct she knew exactly what to do; she started at the inner part of the hind legs, where the skin is softest, then tore out the guts, and after enjoying these delicacies, buried the stomach contents and covered up the blood spoor, as all proper lions do. Then she gnawed the meat off the bones with her molars and rasped it away with her rough tongue.

Once we knew that she could do this it was time for us to let her do her own killing. The plain was covered with isolated bush clusters, ideal hideouts for any animal. All the lions had to do, when they wanted a meal, was to wait under cover until an antelope approached down-wind, rush out and get their dinner.

We now left Elsa alone for two or three days at a time, hoping that hunger would make her kill. But when we came back we always found her waiting for us and hungry. It was heart-breaking having to stick to our programme, when obviously all she wanted was to be with us and sure of our affection. This she showed very clearly by sucking my thumbs and holding on to us with her paws. All the same we knew that for her good we must persevere.

By now we realised that it was going to take us much longer to release her to nature than we had expected; we therefore asked the Government if we could use our long leave in the country for the purpose of carrying out this experiment and, very kindly, they consented. After receiving this permission we felt much relieved since we knew that we should now have the time required for our task.

We increased the number of days on which Elsa was left on her own and we reinforced the thorn fences round our tents, so that they were strong enough to keep any lion out. This we did specifically to prevent Elsa from visiting us when she was hungry.

One morning, when she was with us, we located a lion, who seemed placid and in a good mood; she stepped off the car and we tactfully left the pair alone. That evening while sitting in our thorn-protected tent, we suddenly heard Elsa's miaow and before we could stop her, she crept through the thorns and settled down with us. She was bleeding from claw marks and had walked eight miles back, obviously preferring our company to that of the lion.

The next time we took her a longer distance away from camp.

As we drove we saw two eland bulls, each weighing about 1,500 lb., engaged in a fight. Elsa promptly jumped off the car and stalked them. At first, they were so engaged in their fight that they did not notice her, but when they became aware of her presence she narrowly missed a savage kick from one of them. They broke off the fight and Elsa chased them a short distance and finally came back very proud of herself.

Soon afterwards we met two young lions sitting on the grass in the open. They looked to us ideal companions for Elsa, but by now she was very suspicious of our tricks and would not leave the car, although she talked very agitatedly to them; as we had no means of dropping her off we had to miss this opportunity and went on until we met two Thomson's gazelles fighting; this sight caused Elsa to jump off and we drove quickly away, leaving her to learn more about wild life.

It was nearly a week before we returned. We found her waiting, and very hungry. She was full of affection, we had deceived her so often, broken faith with her, done so much to destroy her trust in us, yet she remained loyal. We dropped some meat which we had brought with us and she immediately started to eat it. Suddenly we heard unmistakable growls and soon saw two lions trotting fast towards us. They were obviously on the hunt and probably they had scented the meat; they approached very quickly. Poor Elsa took in the situation and bolted as hurriedly as she could, leaving her precious meal. At once a little jackal appeared, that up till now must have been hiding in the grass; he lost no time in taking his chance and began to take bite after bite at Elsa's meat, knowing that his luck was not going to last long. This proved true for one of the lions advanced steadily upon him, uttering threatening growls. But meat was meat and the little jackal was not to be easily frightened away; he held on to his possession and took as many bites as he could until the lion was practically on top of him. Even then, with unbelievable pluck he tried to save his meal. Elsa watched this scene from a distance and saw her first meal, after so many days, being taken away from her. In the circumstances it seemed hard that the two lions took no interest in anything but their food and completely ignored her. To compensate her for her disappointment we took her away.

Index

(Page numbers printed in italics refer to captions and illustrations)